Automated Inspection and Quality Assurance

QUALITY AND RELIABILITY

A Series Edited by

Edward G. Schilling

Center for Quality and Applied Statistics
Rochester Institute of Technology
Rochester, New York

1. Designing for Minimal Maintenance Expense: The Practical Application of Reliability and Maintainability, *Marvin A. Moss*

2. Quality Control for Profit, Second Edition, Revised and Expanded, *Ronald H. Lester, Norbert L. Enrick, and Harry E. Mottley, Jr.*

3. QCPAC: Statistical Quality Control on the IBM PC, *Steven M. Zimmerman and Leo M. Conrad*

4. Quality by Experimental Design, *Thomas B. Barker*

5. Applications of Quality Control in the Service Industry, *A. C. Rosander*

6. Integrated Product Testing and Evaluating: A Systems Approach to Improve Reliability and Quality, Revised Edition, *Harold L. Gilmore and Herbert C. Schwartz*

7. Quality Management Handbook, *edited by Loren Walsh, Ralph Wurster, and Raymond J. Kimber*

8. Statistical Process Control: A Guide for Implementation, *Roger W. Berger and Thomas Hart*

9. Quality Circles: Selected Readings, *edited by Roger W. Berger and David L. Shores*

10. Quality and Productivity for Bankers and Financial Managers, *William J. Latzko*

11. Poor-Quality Cost, *H. James Harrington*

12. Human Resources Management, *edited by Jill P. Kern, John J. Riley, and Louis N. Jones*

13. The Good and the Bad News About Quality, *Edward M. Schrock and Henry L. Lefevre*

14. Engineering Design for Producibility and Reliability, *John W. Priest*

15. Statistical Process Control in Automated Manufacturing, *J. Bert Keats and Norma Faris Hubele*

16. Automated Inspection and Quality Assurance, *Stanley L. Robinson and Richard K. Miller*

Other volumes in preparation

Defect Prevention: Use of Simple Statistical Tools, *Victor E. Kane*

Automated Inspection and Quality Assurance

Stanley L. Robinson
Johnson & Johnson International
New Brunswick, New Jersey

Richard K. Miller
Madison, Georgia

MARCEL DEKKER, INC.   New York and Basel
ASQC QUALITY PRESS   Milwaukee

Library of Congress Cataloging-in-Publication Data

Robinson, Stanley L.
    Automated inspection and quality assurance / Stanley L. Robinson,
Richard K. Miller.
        p.   cm. -- (Quality and reliability; 16)
    Includes bibliographies and index.
    ISBN 0-8247-8002-7
    1. Engineering inspection--Automation. I. Miller, Richard
Kendall. II. Title. III. Series.
TS156.2.R62 1989
670.42'5--dc19   88-22699
                                                              CIP

This book is printed on acid-free paper.

MARCEL DEKKER, INC.
270 Madison Avenue, New York, New York 10016

Current printing (last digit):
10 9 8 7 6 5 4 3 2 1

PRINTED IN THE UNITED STATES OF AMERICA

About the Series

The genesis of modern methods of quality and reliability will be found in a simple memo dated May 16, 1924, in which Walter A. Shewhart proposed the control chart for the analysis of inspection data. This led to a broadening of the concept of inspection from emphasis on detection and correction of defective material to control of quality through analysis and prevention of quality problems. Subsequent concern for product performance in the hands of the user stimulated development of the systems and techniques of reliability. Emphasis on the consumer as the ultimate judge of quality serves as the catalyst to bring about the integration of the methodology of quality with that of reliability. Thus, the innovations that came out of the control chart spawned a philosophy of control of quality and reliability that has come to include not only the methodology of the statistical sciences and engineering, but also the use of appropriate management methods together with various motivational procedures in a concerted effort dedicated to quality improvement.

This series is intended to provide a vehicle to foster interaction of the elements of the modern approach to quality, including statistical applications, quality and reliability engineering, management, and motivational aspects. It is a forum in which the subject matter of these various areas can be brought together to allow for effective integration of appropriate techniques. This will promote the true benefit of each, which can be achieved only through their interaction. In this sense, the whole of quality and reliability is greater than the sum of its parts, as each element augments the others.

The contributors to this series have been encouraged to discuss fundamental concepts as well as methodology, technology, and procedures at the leading edge of the discipline. Thus, new concepts are placed in proper perspective in these evolving disciplines. The series is intended for those in manufacturing, engineering, and marketing and management, as well as the consuming public, all of whom have an interest and stake in the improvement and maintenance of quality and reliability in the products and services that are the lifeblood of the economic system.

The modern approach to quality and reliability concerns excellence: excellence when the product is designed, excellence when the product is made, excellence as the product is used, and excellence throughout its lifetime. But excellence does not result without effort, and products and services of superior quality and reliability require an appropriate combination of statistical, engineering, management, and motivational effort. This effort can be directed for maximum benefit only in light of timely knowledge of approaches and methods that have been developed and are available in these areas of expertise. Within the volumes of this series, the reader will find the means to create, control, correct, and improve quality and reliability in ways that are cost effective, that enhance productivity, and that create a motivational atmosphere that is harmonious and constructive. It is dedicated to that end and to the readers whose study of quality and reliability will lead to greater understanding of their products, their processes, their workplaces, and themselves.

Edward G. Schilling

Preface

The assurance of product quality is one of the most important goals of every industry, both within the United States and internationally. Assuming our readers are already convinced of the importance and paybacks of quality assurance, this book does not argue the importance of quality control, but rather addresses techniques for applying state-of-the-art technologies to achieve an effective quality assurance program in automated manufacturing. Specifically, quality assurance is viewed in the context of industrial automation and computer-integrated manufacturing.

In the evolution of automated quality assurance, two points are important to recognize. First, the implementation of automated quality assurance preceded computer use on the shop floor. As far back as the 1960s, we saw the use of a wide variety of mechanical, electrical, and optical systems for automated quality control. Second, the introduction of the computer per se did not immediately foster the current generation of automated quality assurance technology or computer integrated manufacturing. The computer era began with the mainframe, which was not a real-time system. It was the "microprocessor revolution," not the "computer revolution," that gave rise to computer-integrated manufacturing, had revolutionary implications for quality assurance, and is the focus of this book. Consider the following dollar value distribution of the computer market:

Year	Total market ($B)	Main frames (%)	Mini-computer (%)	Micro-computer (%)	Other (%)
1975	13	83	10	0	7
1980	29	60	17	6	17
1985	63	36	21	20	23
1990	141	17	21	37	25
1995	313	6	16	55	23

Real-time, shop-floor, computer-based systems for quality control were the focus of considerable research in the 1970s, and some engineers foresaw "the factory of the future." In the early 1980s, industry saw the widespread use of microcomputers and microprocessor-based, application-specific computer systems, such as machine vision. Currently, the microcomputer-on-a-chip is becoming the dominant form of computer. A large portion of the microcomputers-on-a-chip are being put into and under the covers of other products. This is often referred to as the "brain in the machine." Whether that machine brain is embedded with conventional algorithms (e.g., Fortran C) or artificial intelligence (e.g., knowledge-based systems), it is being programmed to serve as a tool to achieve quality in manufacturing.

This book benefits from the engineers and product manufacturers who have developed the systems that fostered the revolution in automated quality assurance. Many have contributed background information for this book. To all we are indebted for the material contained here and to each we extend our appreciation.

Stanley L. Robinson
Richard K. Miller

Contents

About the Series   iii

Preface   v

1 HISTORICAL OVERVIEW   1

**2 HOW AUTOMATED QUALITY ASSURANCE
AFFECTS THE WORLD   7**

   2.1   Characteristics of the Old World   7
   2.2   Characteristics of the New World   9
   2.3   Wheel of Progress   11

3 COMPUTER-INTEGRATED MANUFACTURING   15

   3.1   Computer-Integrated Manufacturing   17
   3.2   How Computer-Integrated Manufacturing Works   20
   3.3   Computer-Integrated Manufacturing and
        Automated Quality Assurance   25
   3.4   Just-In-Time Manufacturing   25

3.5	Material Resource Planning	27
3.6	Computer-Aided Process Planning	30
	Bibliography	32

4   MACHINE VISION — 33

4.1	Solid-State Cameras	34
4.2	Resolution	36
4.3	Lighting	36
4.4	Connectivity Analysis	39
4.5	Binary Processing and Gray Scale	43
4.6	Three-Dimensional Vision	43
4.7	Commercial Machine Vision Systems	45
4.8	Future Trends	46
	Bibliography	48

5   SENSORS FOR INDUSTRIAL INSPECTION — 49

5.1	Introduction	49
5.2	Ultrasonic Sensors for Automated Inspection	56
5.3	Automatic Inspection for Packaged Products	65

6   ROBOTIC INSPECTION — 71

6.1	Servo Robots	72
6.2	Fundamentals of Robot Control and Programming	72
6.3	Types of Robots	74
6.4	Reliability	76
6.5	Intelligent Robots	76
6.6	Robot Vision	76
6.7	Robotic Testing and Inspection	77
6.8	Automobile Body Gauging	80
6.9	TCM Board Tester	82
6.10	Valve Testing	82
6.11	X-Ray Testing	85
6.12	Detector Alarm Check	86
6.13	Electrical Contact Testing	88

7   SOFTWARE FOR QUALITY ASSURANCE — 91

7.1	Operating System	92
7.2	How to Select Software for Automated Quality Assurance	93

7.3   Verifying Software Reliability   95
7.4   Directory of Automated Quality Assurance
      Software   96
      Bibliography   96

8   INSPECTING THE PRODUCT   99

8.1   Vision Inspection at General Motors   100
8.2   Vision Inspection at Kodak   100
8.3   Vision Inspection at Saab   101
8.4   Sheet Metal Inspection   102
8.5   Automatic Inspection of Engine Blocks   108
8.6   Pharmaceutical Inspection   111
8.7   Quality Assurance for Semiconductors   114
8.8   Automated Printed Circuit Board Inspection   115
8.9   Nonwoven Materials Inspection   121
8.10 Casting Flaw Detection   123
8.11 Glass Tubing Inspection   124
8.12 Other Applications   124
      Bibliography   125

9   INSPECTING THE PACKAGE   127

9.1   Container and Label Inspection   128
9.2   Alphanumeric Character Inspection   131
9.3   Automated Closure Inspection   134
9.4   Pharmaceutical Product Counting   136

10 DEVELOPING THE IN-HOUSE PROGRAM   139

10.1   The Automated Quality Assurance Plan   140
10.2   Implementing Machine Vision   141
10.3   Implementing Robotic Inspection   143
10.4   The Computer Communications Problem   146
10.5   Manufacturing Automated Protocol   147
10.6   The Technology of Manufacturing
       Automated Protocol   148
10.7   Other Computer Protocols and Standards   151
10.8   Computer Simulation   153
10.9   Simulating a Robotic Inspection Work Cell   155
10.10 How to Simulate an Automated Quality
       Assurance Project   158
10.11 Assessing the Benefits   159

11 ARTIFICIAL INTELLIGENCE: THE NEXT STEP — 163

11.1 Expert and Knowledge-Based Systems — 163
11.2 Expert Systems for Quality Assurance — 167
11.3 The Technology of Expert Systems — 171
11.4 Tools for Building Expert Systems — 182
      Bibliography — 182

12 CASE STUDIES — 185

12.1   Allen-Bradley — 186
12.2   AT&T — 189
12.3   Deere & Company — 190
12.4   Frost, Inc. — 192
12.5   General Electric — 193
12.6   General Motors — 198
12.7   Merck & Company, Inc. — 200
12.8   Texas Instruments — 202
12.9   3M — 204
12.10 Whirlpool — 207

Appendix 1: Glossary — 213

Appendix 2: Commercially Available Software — 231

Index — 253

1

Historical Overview

Quality in manufacturing has always been of interest to American industry. It has only been within the past decade, however, that computers have been applied to automating the quality assurance task. The batch processing of mainframe computers of the pre-1970 era was not applicable to real-time operation, and these computers were much too costly for dedicated industrial operations. When the development of minicomputers opened the door to manufacturing applications of computers, inspection was one of the first tasks considered. The minicomputers of the early 1970s offered only a fraction of the capabilities of today's personal computers at several times the cost. However, they were sufficient for researchers to begin to develop the first machine vision systems and other technologies which today form the basis of automated quality assurance. This chapter reflects on some of these early efforts which matured to a wide range of commercial inspection products in the late 1970s and early 1980s.

The National Science Foundation (NSF) undertook a Production Research and Technology program between 1973 and 1982. The objective of this program was to develop a new technology called "programmable automation" to advance automation in batch manufacturing of discrete products for both economic and social reasons. Two of the primary technologies developed were machine vision and robotics. Among the primary contractors were Stanford University, SRI International, Purdue University, and the University of Rhode Island.

The NSF sponsored program at SRI International, begun in April 1973, was the driving force behind the commercialization of machine vision systems nearly a decade later. The research utilized a 128 × 128 solid-state camera and a DEC LSI-11 microcomputer with a 28K word memory. While these components offered only a fraction of the capabilities of today's commercial systems, SRI developed some important techniques. A software routine called "connectivity analysis" allowed the analysis of objects (e.g., dimensions, area, etc.) and the extraction of local features (e.g., holes and corners). A method was developed for using gray scale to inspect registered objects for pseudocolor. Structured light was applied to inspection of three-dimensional objects by comparing model lines with the inspected lines of intersection of the light plane and indexed three-dimensional objects. All of these techniques of now in popular use, achieving production line speed capabilities due to advance in (VLSI) technology in the decade following the SRI research.

The industrial robot was invented in 1951 by George C. Devol, and commercialized in 1961 by Joseph Engleberger and his company, Unimation. Early industrial robots, however, were controlled by mechanical switches rather than computers. They were heavy-duty machines designed for transporting large loads in the harsh environments of foundries, forges, and metal fabrication plants. The era of computer-controlled "intelligent" robots, applicable to such sophisticated tasks as inspection and assembly, had its roots in university laboratories, primarily sponsored by the National Science Foundation program.

The first program in robotics and computer vision was started at Stanford University in 1965 by John McCarthy. Several landmark accomplishments in the robotics field were recorded in the early years of the program at Stanford. The Scheinman Stanford robot arm (1970) was the forerunner of modern robots used for industrial assembly. The WAVE programming system (1971), developed by Richard Paul, was the first to provide several important capabilities of modern robot programming systems: predictive Newtonian dynamics, automatic planning of smooth trajectories, rudimentary force and touch sensing used in control, and a macro library of assembly operations. This led to the first computer-integrated assembly in 1973, the assembly of an automotive water pump from 10 component parts. By the late 1970s, the Scheinman arm had evolved into the commercial PUMA robot, and WAVE evolved into the commercial VAL programming system. The age of industrial robotics had begun.

Some important contributions to robotics were also developed under NSF sponsorship by Dr. R.B. Kelley's group at the University of Rhode Island. Studies centered on methods to enable robots with vision to acquire, orient, and transport workpieces. A method was developed to allow a vision guided robot to acquire randomly placed workpieces from a bin. This "bin picking problem" was listed first among the frustrating gaps in needed knowledge for factory automation, according to a 1976 poll of members of the Society of Manufacturing Engineers. Also developed in Kelley's program was an instrumented parallel jaw gripper which allowed a robot to "feel" the workpiece within its grip.

The NSF sponsored the development of advanced industrial robot control systems at Purdue University, under the direction of Richard Paul and Shimon Nof. Among the accomplishments were advanced position, velocity, and force controls, and a technique for simulating a robots work routine. Simulation subsequently has become a routine practice for engineers in developing robotic work cells.

The application of this research to automated quality assurance and computer-integrated manufacturing began in 1978. In that year, Machine Intelligence Corporation, founded by researchers from SRI International, introduced the first machine vision used for inspection. Also in 1978, General Motors demonstrated the CONSIGHT vision system, used in conjunction with a Cincinnati Milacron robot for sorting metal castings.

The November 1981 AUTOFACT conference in Philadelphia is recognized as the event which launched the widespread recognition of machine vision technology. At this event over a dozen companies demonstrated commercial vision inspection systems and over 5,000 engineers left with the knowledge of how this technology could be utilized for automated quality assurance within their companies. The following year saw IBM, Control Automation, Intelledex, and other companies introduce robot models with extreme precision and computer vision capabilities for sophisticated inspection and assembly tasks. Automatix was recognized for the marriage of robotics with a statistical process control software package. Lord Corporation and Barry Wright Corporation introduced advanced tactile sensor pads for robot grippers. The computer-based techniques of machine vision were expanded to other types of sensors for inspection. Based on research at MIT, Cochlea Corporation developed a commercial computer-based ultrasonic system for the inspection of small parts.

Today, machine vision systems for inspection and industrial robots have become commonplace. Over 10,000 computer vision systems and over 20,000 industrial robots have been installed in industries of the United States. But this is only a beginning. Of the 700,000 different quality control tests run regularly in the United States, it is estimated that at least 25% could be replaced by fully automated machine vision inspections. An additional 40% could be more effectively handled by an operator using machine vision as a gauge. A report by the National Bureau of Standards suggested that on the order of 90% of all industrial inspection activities requiring vision will be done with computer vision systems within the next decade.

The potential cost savings achievable by automating the repetitive, boring inspection task through the use of automated inspection is tremendous. Inspection tasks are performed by an estimated 10% of the U.S. workforce. This translates to 400,000 persons. If an average annual wage of $20,000 is assumed, the annual cost of industrial inspection in the United States is $8 billion. The incentive for U.S. industry to implement automated quality assurance is not only to turn this cost into profit through automation, but also to increase product quality as necessary for U.S. industry to maintain its position as world leader in manufactured goods.

2

How Automated Quality
Assurance Affects the World

2.1   Characteristics of the Old World
2.2   Characteristics of the New World
2.3   Wheel of Progress

In order to evaluate the effect of automated quality assurance (AQA) on the world, we will compare two worlds. The present predominant world will be referred to as the "Old World." The after-automation world is the "New World."

2.1 CHARACTERISTICS OF THE OLD WORLD

The Old World, before the influence of automated quality assurance, has the following characteristics:

1. A fully-manned, costly, quality control department: managers, supervision, on-line inspectors, sampling inspectors, and data handlers.

2. Use of sampling techniques concerned with outgoing quality level (OQL) which yield a less-than-perfect inventory of finished product.

3. The human on-line inspectors become fatigued and perform in less-than-satisfactory manner.

4. Often destructive sampling is essential, which results in a costly quality control method.

5. Scrap loss is elusive. The tendency is to use broad stroke methodologies, i.e., withdrawals of raw material from warehouse equals input. Finished product equals output. The difference is scrap loss. Not necessarily true, i.e., inaccurate warehouse records, material moisture control, etc.

6. After-the-fact control samples taken and inspections made followed by time-consuming analysis while good or bad product continues to be made.

7. Simple manual data processing is used for record-keeping. This limits the degree and extent of data collection. Process of record-keeping is relatively slow and costly.

8. Manufacturing is not a real-time process. To know what is happening while it happens allows for a better decision-making process.

9. If raw material varies, the process no doubt varies. The standard deviation increases without the knowledge of the operators. It can happen in small increments that are not detectable by human eye.

10. Output can go beyond the upper control limits (UCL) or lower control limits (LCL) without human detection.

11. Process control is only a broad stroke method. Data collected by sampling methods is general, nonspecific, and not timely. Process could move in or out of the control limits before knowledge of sample results are available.

12. A process is practical when it has a known margin of error, i.e., you are allowed to deviate from the mean and still produce an acceptable product. Acceptable product may mean a product with

some compromises in order to make the process usable and practical. The less the knowledge about the process data, the greater the potential for compromises.

13. The Old World deals in macrocosms. It evaluates end results, and is not capable of finer scrutiny.

14. System (old) is not on-line, real time, user-friendly.

> Slow to conclude process adjustments
> Slow scrap control
> Does not evaluate supplier

15. Specification limits are usually broad due to lack of process insights through inadequate information.

16. Tendency to allow higher work in process to cover for contingency.

17. Higher risk for back order due to less control and predictability.

2.2 CHARACTERISTICS OF THE NEW WORLD

The New World—after the introduction of automated quality assurance—may be characterized as follows:

1. AQA allows better evaluation of suppliers which can lead to just-in-time (JIT). The results are increased inventory turnover in magnitudes of a present situation of four to six increasing to 20 to 40. Consequently work-in-process (WIP) is reduced.

2. One-hundred percent inspection becomes possible. Slow untimely sampling procedures are replaced with high-speed accurate information. Outgoing quality level approaches 100% acceptable product.

3. Analysis in microcosms becomes possible. Inspection points can be increased (e.g., inspection can occur at several points of a machine as opposed to one human inspection at the discharge point).

4. Reduced quality control manpower results in a cost reduction.

5. On-line test of raw materials is possible at points of entry to the process. Supplier evaluations are a convenient result.

6. Automated quality assurance is superior to human effort—less fatigue and less variance.

7. Increased points of inspection before the final product is assembled can eliminate costly after-the-fact destructive inspections.

8. Scrap avoidance can be a built-in factor. High-speed data collection in real time at a greater number of points in the process allows for better decisions.

9. Before-the-fact control or synchronous control becomes possible.

10. Computer integration with other segments of a work cell or within a department becomes possible.

11. Automated quality assurance is real time, user-friendly.

12. Process variation can be tracked in smaller increments or continuously with speedy data collection and interpretation resulting in usable information.

13. Process control takes on a new profile of technology. Managers know what is really happening as opposed to approximation.

14. Specification limits do not have to be as broad. Product quality is improved.

15. The practical margin of error can be identified as well as reduced with AQA.

16. Customer satisfaction should increase due to a better, more consistently performing product.

17. Automated quality assurance can see better than the human eye.

18. Better control of the process requires less work-in-process (WIP).

19. Less work-in-process results in reduced space requirements. Also, quality control offices, area, and inspection stations.

20. Planning and scheduling improve by being quicker, better due to previously unavailable information.

21. Management systems improve and are simplified—less critical management, more logic in the work environment, greater predictability, better equipped to deal with the unexpected.

2.3 WHEEL OF PROGRESS

The new information base made possible by automated quality assurance raises the odds of utilizing computer-integrated manufacturing (CIM) in an efficient manner. Real-time dependable process control through AQA yields intelligent information that can serve to drive other disciplines in the business.

In order to compete as a world class manufacturer (WCM), a company must build a reliable production system beyond Old World imagination. Automated quality assurance serves as that foundation and is the stepping-off point to many other techniques characteristic of a world class manufacturer. For example:

CAD — Computer-Aided Design
CAE — Computer-Aided Engineering
NC — Numerical Control
MRP — Manufacturing Resource planning

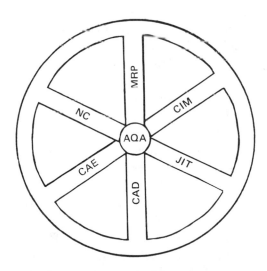

Fig. 2.1   The wheel of progress.

CIM — Computer Integrated Manufacturing
JIT   — Just-In-Time

Automated quality assurance is the hub of the wheel (Figure 2.1) and the spokes are made of the various techniques that depend on a reliable, well-controlled production system.

A major problem in the evolution of a world class manufacturer is the lack of a method of learning for the mass of practitioners. The need for perpetual education is the key issue. The technological explosion is increasing at a meteoric rate. If education is not in a perpetual mode, the practitioner quickly finds himself outdated. If he avoids involvement in the perpetual education process, he will never achieve world class manufacturer status.

Automated quality assurance applies to this decision in a significant way. Total fulfillment as a world class manufacturer has as an essential ingredient AQA as the hub of the wheel of progress.

The primary product costs are labor, material, and overhead. A conventional breakdown in order of magnitude is:

Material	75%
Overhead	25%
Labor	5%

The current trends show material costs continuing to increase as a piece of the pie. Automated quality assurance (AQA) directs itself concentratively to the management of the material resource. Scrap control, material utilization, and supplier evaluation are automatic built-ins in the AQA system. It is further worthwhile repeating that without AQA many of the other technologies of WCM are impossible. JIT cannot happen without a total quality control commitment. Automation and CIM are coupled with the principle of AQA. Once again, AQA is the hub of the wheel of progress towards manufacturing excellence.

The factory of the future (FOF) may be described as a Greenfield Site (new proposition) where a "Lights Out" plant (unmanned) is developed utilizing the Steel Collar Worker (robots). This factory of the future takes full advantage of world class manufacturer technologies such as CIM, MRP, NC, CAD, CAE, and of particular note is automated quality assurance.

The exportation of U.S. manufacturing must cease. Off shore avenues become less attractive under the WCM principle. Foreign competition will be less viable when we create our own competitive advantage in manufacturing. Lead time, product cost, flexibility, and most of all, superior quality can only come about through a production system that is primary in a global rank. It only remains to say that tantamount to all of this is a system of quality assurance that keeps pace with the meteoric technological explosion, and the fundamentals of the factory of the future. Needless to say, the solution is the unique, reliable, real-time, user-friendly technology of the future, automated quality assurance, a major competitive advantage in itself.

3

Computer-Integrated Manufacturing

3.1   Computer-Integrated Manufacturing
3.2   How Computer-Integrated Manufacturing Works
3.3   Computer-Integrated Manufacturing and
       Automated Quality Assurance
3.4   Just-In-Time Manufacturing
3.5   Material Resource Planning
3.6   Computer-Aided Process Planning
       Bibliography

The computer-integrated manufacturing (CIM) factory of the future (FOF) will look quite different from the factory of today. It will be based upon the integration of the traditional, process-based factory technology of today with the emerging software-systems-based technology of today and tomorrow (see Figure 3.1). The factory of the eighties must combine the following characteristics:

15

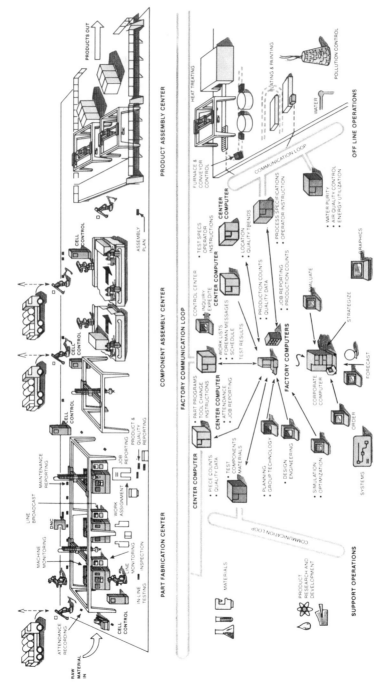

Fig. 3.1 Factory of the future concept. (Courtesy of GCA Corporation.)

16

Efficiently integrated and continuous operations
Flexible and economical in the face of change
Effectively organized for maximum productivity and satisfied
  employees
Produce a quality and cost-competitive product

Recently the National Aeronautics and Space Administration (NASA) asked the National Research Council (NRC) to study the impact of integration efforts at McDonnell Aircraft Co., Deere and Co., Westinghouse Defense and Electronics Center, General Motors, and Ingersoll Milling Machine Co. The NRC's Committee on the CAD/CAM Interface checked and rechecked the results carefully before printing them. The committee found these companies had already received significant benefits even though they were only partially into their 10- to 20-year integration efforts. Benefits identified are:

Reduction in engineering design cost:	15-30%
Reduction in overall lead time:	30-60%
Increased productivity of production operations (complete assemblies):	40-70%
Reduction of work in process:	30-60%
Increased product quality as measured by yield of acceptable product:	2-5 times
Increased capability of engineers as measured by extent and depth of analysis in same or less time than previously:	3-35 times
Increased productivity (operating time) of capital equipment:	2-3 times

3.1 COMPUTER-INTEGRATED MANUFACTURING

Computer-integrated manufacturing (CIM, pronounced "sim") involves the integration and coordination of design, manufacturing, and management using computer-based systems. Computer-

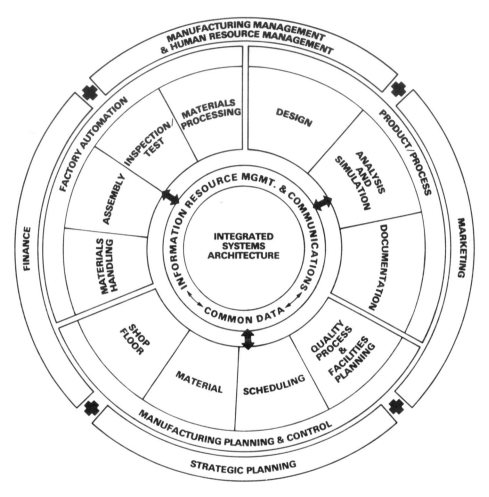

Fig. 3.2   The CIM wheel. (Courtesy of CASA/SME; copyright, 1985.)

integrated manufacturing is not yet a specific technology that can be purchased, but rather an approach to factory organization and management.

The "CIM Wheel," Figure 3.2, developed by the Technical Council of the Computer and Automated Systems Association of SME, shows the elements of CIM.

Computer-integrated manufacturing was first popularized by Joseph Harrington's book of the same name, published in 1974. One systems expert recounts the history of the concept in this way:

[CIM] came about from: (1) The realization that in many cases automation for discrete activities in manufacturing, such as design or machining, in fact often decreased the effectiveness of the entire operation—e.g., designers could conceive parts with CAD that could not be made in the factory; numerical control (NC) machine tools required such elaborate setup that they could not be economically programmed or used. (2) Development of large mainframe computers supported by data base management systems (DBMS) and communications capabilities with other computers. The DBMS and communications allowed functional areas to share information with one another on demand. (3) The dawning of the microcomputer age which began to allow machines in the factory to be remotely programmed, to talk to each other and to report their activity to their ultimate source of instruction.

Though there is no quantitative measure of integration in a factory, and definitions of CIM vary widely, the concept has become a lightning rod for technologists and industrialists seeking to increase productivity and exploit the computer in manufacturing. For example, James Lardner, vice president of Deere & Co., sees the current state-of-the-art manufacturing process as a series of "islands of automation," in which machines perform tasks essentially automatically, connected by "human bridges." The ultimate step, he argues, is to connect those islands into an integrated whole through CIM and artificial intelligence, replacing the human bridges with machines. In this essentially "unmanned factory," humans would then perform only the tasks that require creativity, primarily those of conceptual design. Lardner's vision is echoed by many other prominent experts.

Experts differ in their assessment of how long it might take to achieve this vision—virtually no one believes that it is attainable in less than 10 to 15 years, while some experts would say an unmanned factory is at least three decades away. More importantly, there are other technologists who argue that the vision may, in fact, be just a dream. For example, Bernard Roth, professor of mechanical engineering at Stanford University, argues that factories will, in reality, reach an appropriate and economical level of automation and then the trend toward automation will level off. In a sense, the difference between these two views may be a difference of degree rather than kind. For many factories, the "appropriate" level of automation might indeed be very high. In others, however, a fair number of humans will remain, though they may be significantly fewer than currently present.

Integrated systems are often found to require more human input than was expected. Indeed, as one engineer explains:

> There is much talk about the totally automated factory—the factory of the future—and night shifts where robots operate the factory. Whereas the situations will develop in some cases . . . many manufacturing facilities will not be fully automated. Even those that will involve humans in system design, control, and maintenance—and the factory will operate within a corporate organization of managers and planners.

These two views do have important significance for how an industrialist might now proceed. Many who hold the vision of the unmanned factory seem to emphasize technologies, such as robotics, that can remove humans from manufacturing. Those who do not share the vision of "unmanned manufacturing" tend to argue that there are more practical ways to enhance productivity in manufacturing, including redesigning products for ease of fabrication and assembly.

3.2 HOW COMPUTER-INTEGRATED MANUFACTURING WORKS

There are two different schemes for computer-integrated manufacturing (CIM): In vertically integrated manufacturing, a designer would design a product using a computer-aided design (CAD) system, which would then translate the design into instructions for production on computer-aided manufacturing (CAM) equipment. Management information systems and computer-aided planning systems would be used to control and monitor the process. A horizontal approach to integration, on the other hand, would attempt to coordinate only the manufacturing portion of the process; i.e., a set of computer-aided manufacturing equipment on the factory floor is tied together and coordinated by computer instructions. A flexible manufacturing system would be a good example of such horizontal integration. Vertically integrated manufacturing is what is most commonly meant by CIM, however, and many experts would consider horizontal or "shop floor" integration to be only partial CIM. Figure 3.3 is a conceptual framework for CIM which illustrates the role of some of the programmable automation technologies at various levels of factory control.

A vertically integrated factory usually implies maximum use and coordination of all programmable automation technologies,

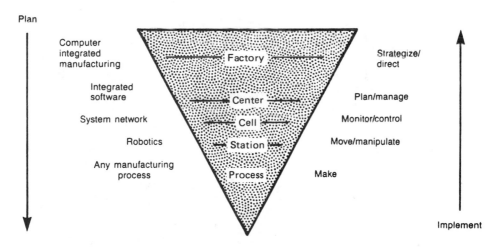

Fig. 3.3 Programmable automation factory hierarchy. (Courtesy of GCA Corporation.)

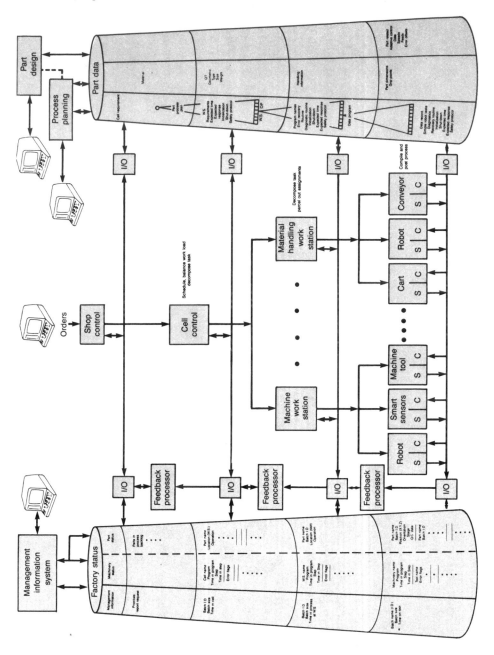

and can involve much more centralized control of manufacturing processes than a nonintegrated production process. Communication and shared data bases are especially important for CIM. For example, CAD systems must be able to access data from inventory on the cost of raw materials, and from CAM systems on how to adapt the design to facilitate manufacture. Computer-aided manufacturing systems must be able to interpret the CAD design and establish efficient process plans. And management computer tools should be able to derive up-to-date summary and performance information from both CAD and CAM data bases, and effectively help manage the manufacturing operation.

Some parts of the above requirements are already possible, while others seem far on the horizon. Factory data bases now tend to be completely separate, with very different structures to serve different needs. In particular, the extensive communications between CAD and CAM data bases will require more sophistication in both CAD and CAM, research on how to establish such communications, and finally, major changes in traditional factory data structures in order to implement such a system.

One of the key issues in CIM development involves the logistics of a complicated factory. Several groups, including the National Bureau of Standards (NBS), the U.S. Air Force ICAM project, and Computer-Aided Manufacturing International (CAM-I), have been working on "architectures" for such an automated factory. Figure 3.4 is an example of such a conceptual framework for CIM which forms the foundation for detailed work on factory control architectures.

One of NBS's major contributions in automation research and development has been in developing strategies for the interface of programmable automation devices. Their emphasis has been on what they call a "mailbox" or decentralized approach to factory communication and control. In such a system, the control of the

---

Fig. 3.4   National Bureau of Standards scheme for distributed factory control.

factory is distributed at different levels among the various programmable automation devices. For example, a factory-level computer might send a message to a production-level computer: "Make 150 of part number 302570." The production-level computer would then send a message to the "mailbox" of a certain work cell: "Execute production plan for part 302570, 150 times." In turn, the work cell controller would send messages to the mailbox of the machine tools and robots in the cell, to execute certain programs stored in their memory.

The "mailbox" approach differs from a centralized, or "star," approach to automated systems control in which a central computer directly controls each action of every machine in the factory. The advantages of the mailbox system are that it simplifies standards and interface problems—the only interface standard necessary is for the location of the mailbox in which to deposit messages. For example, this allows one robot to be substituted for another with relative ease. The mailbox approach allows different programmable automation devices to operate using different languages and proprietary operating systems, as long as they are able to interpret messages from the computer controller.

Hierarchical arrangements for automated manufacturing tend to involve a large number of separate computers, each with separate data bases. Techniques for "distributed data base management," that is, managing and manipulating data in several computer systems simultaneously, need to be developed in order for a hierarchical arrangement to be practical. Similarly, techniques and standards for establishing communication between computerized devices, both in-plant and between plants, need to become much more sophisticated.

T. Williams and a group of researchers at Purdue University, in collaboration with several large manufacturers, are attempting to exploit currently available technology to design an actual factory with maximum computer integration. The leader of that effort argues that the technology for CIM is available, and that technical advances, though welcome, are not necessary. Rather, he argues that factors holding back "fully" automated manufacturing are primarily:

The lack of standards for interfaces, communication networks, and programming languages

A need for more powerful data-base management systems

The need for detailed mathematical models of physical and chemical processes

Shortages of technical personnel

Shortages of computer power

Manufacturing management who are unaware of the detailed technical benefits of automation.

3.3 COMPUTER-INTEGRATED MANUFACTURING AND AUTOMATED QUALITY ASSURANCE

Using computer-integrated manufacturing (CIM), potential fit and function problems in the actions and interactions of the manufactured part, and its tooling and assembly procedures, can be successfully investigated sooner and remedial action take earlier.

The resulting data has significant value for statistical quality control. The end product is more easily produced and is less costly to make. Customers get a better product. Computer-integrated manufacturing makes it easier to investigate those "special" manufacturing headaches and develop tooling alternatives before production problems actually occur on the shop floor.

If the product is manufactured to strict design tolerances, CIM will lessen the probability of a tolerance or "fit" problem. Computer-integrated manufacturing can significantly reduce both defects and rework with resulting cost savings and cost avoidance.

3.4 JUST-IN-TIME MANUFACTURING

Just-in-time (JIT) manufacturing is a philosophy that has the elimination of all non-value adding operations, equipment, and resources at its heart. Since its introduction in the United States, JIT has helped manufacturers reduce costs, increase productivity, and challenge foreign competitors once again.

Just-in-time manufacturing simplifies the manufacturing process by eliminating waste throughout the system. This includes inventory at both ends, and all material, machines, and manpower that do not contribute directly to the product. Activities such as moving, storing, counting, sorting, scheduling, and quality inspection, which add cost but not value, are prime targets.

The true strategic objective of JIT is not to decrease inventory, but to increase profit. Inventory will fall, as an effect of JIT, not as an achieved objective.

Most of America's largest and most innovative companies have experimented with or implemented JIT manufacturing to some degree. These include Hewlett-Packard, General Electric, International Business Machines, General Motors, Apple Computer, Chrysler, American Motors, Burroughs, Deere & Co., Ford, Bendix, Hyster, and Digital Equipment Corporation. Smaller companies include Black and Decker, Allis Chalmers, and Harley Davidson, to name a few.

General Motors is implementing JIT at the new Detroit/Hamtramck assembly plant, organizing just-in-time deliveries of 90% of its parts. Few U.S. plants have implemented JIT so fully. 1700 suppliers deliver 6500 parts exactly in the sequence in which they are supposed to be installed for four models. Upon receiving dealer orders, the plant fixes its production schedules at least 10 days in advance. The plant then notifies in-sequence suppliers of its schedule for the next 10 days in four-hour increments. Loading docks receive just-in-time parts at least once per hour, and the plant stores just 2 to 4 hours inventory of the 1700 in-sequence parts. The plant is scheduled to reach full production in late 1986.

Harley Davidson, a JIT pioneer, reduced its set-up time by more than 75% by moving to semicircle-shaped cells; lineal footage was reduced from nearly 35,000 feet to 13,000 feet. Thirty percent of the floor space was freed up. Effective rate of completion on motorcycles went from 65% in 1981 to between 99 and 100 percent today. Turnaround time dropped from three and a half days to less than one-half day. Some 70,000 square feet of stockroom containing spare components was eliminated.

Tractor manufacturer Allis Chalmers reduced set-up time on one machine from 11 hours to less than 4 hours. By combining rough and finish turning departments into one, the company eliminated three forklift operations, three inspection processes and four additional sets of paperwork.

IBM's Lexington, KY, typewriter plant reduced its number of suppliers tenfold. It can manufacture typewriter motors with guaranteed zero defects. By changing the product design for automated manufacturing and JIT, IBM reduced the parts it uses 3:1, reduced the time needed to manufacture 4:1, and decreased average adjustments 9:1.

Sargent & Co. reduced late orders from 62% to 5% on a higher sale volume. The average late order is now only one week late, rather than the 14 weeks late previous to implementing JIT. Inventory takes up 80,000 square feet less than it used to. Today two divisions work in the space previously occupied by one (U.S. Office of Technology Assessment, 1984).

How does a company get started with JIT? The first step successful implementors take is to look for opportunities to improve internal operations. JIT is much more than eliminating multiple suppliers, and having suppliers deliver only parts when needed. To get moving with JIT, start internally, and only when satisfied with results there, look to the outside for further improvements.

3.5 MATERIAL RESOURCE PLANNING

IBM developed the first Material Requirements Planning (MRP) software package in the mid-1960s as a part of the Production Information Control System (PICS). This package is the forerunner of over 140 MRP software packages on the market today. The introduction of that first PICS piece of software marked a turning point in the way manufacturing inventories were managed.

The first material requirements planning packages were primarily inventory ordering systems. Data on what is to be made (the Master Production Schedule), what is required to make it (the Bill of Materials), and the materials on hand (the Inventory Record), were used to calculate what should be obtained (Material Requirements).

With the ability to have valid schedules that material requirements planning provides, people using the system realized that several other fundamental functions were required. Closed-loop material requirements planning evolved, so called because it requires feedback from vendors, the factory, the planners, etc. While based on the simple logic of earlier MRP programs, closed-loop systems included features such as production planning, master production scheduling capabilities, capacity planning, and tools for executing and monitoring both capacity and material plans.

In the late 1970s, still more capabilities were added with the development of manufacturing resource planning. Known as MRP-II, the new system incorporates all the elements of closed-loop MRP, but also integrates financial planning with operational planning, allowing the entire company to manage with the same set of "numbers." It includes simulation which permits management to test various production plans. Simply stated, MRP-II is a set of computer-supported management tools for planning and monitoring all the resources of a manufacturing company more professionally. Modules in the MRP-II software package typically include:

> File management (part number, bill of materials, routings, work centers)
> Master production scheduling
> Inventory control
> Material requirements planning
> Capacity requirements planning
> Manufacturing order control
> Shop floor control
> Production costing (cost accounting)
> Purchasing management
> Accounts payable
> Accounts receivable
> General ledger
> Order entry
> Sales analysis

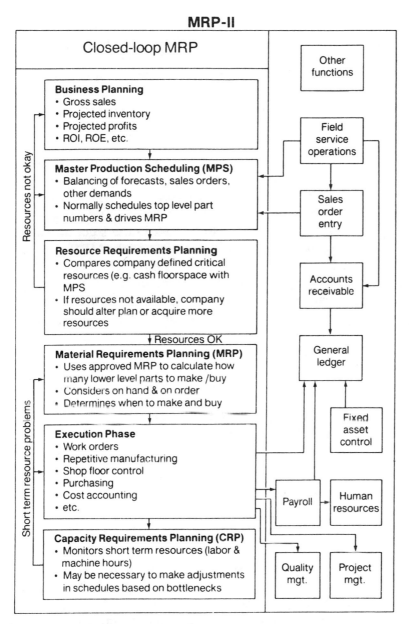

Fig. 3.5 The differences between MRP, closed-loop MRP, and MRP-II systems. (Courtesy of *Production Engineering*, Feb. 1986, pg. 63.)

There is some confusion as to the difference between MRP, closed-loop MRP, and MRP-II systems. Users want a complete MRP-II system, but will find many vendors are in reality only selling MRP packages. Figure 3.5 will help in understanding what features are needed in an MRP-II package.

The installed base of MRP-II systems in the United States passed the 25,000 mark in 1985, and penetration has reached 25% in certain industries.

The American Production and Inventory Control Society (APICS) has categorized companies that have attempted to implement MRP-II into Class A, Class B, Class C and Class D users. Class A users are characterized as organizations that get full anticipated benefits from the system, while Class D users get no more than 50% of anticipated benefits. Some 10% of the MRP-II users can be classified as Class A, according to the APICS. The number of companies that fall into Class D is approximately 27%, and the remaining 63% can be categorized as Class B and Class C users. While Class B and Class C users have some problems with the system, they do receive some anticipated benefits from it and are not considered disasters.

The potential benefits in an effective, well-planned implementation of MRP-II are enormous:

Raw material inventory reduction: 13-40%
WIP reduction: 10-40%
Through-put increase: 50-100%
Customer service levels achieved: 98-100%

3.6 COMPUTER-AIDED PROCESS PLANNING

A computer-aided planning tool is computer-aided process planning (CAPP), used by production planners to establish the optimal sequence of production operations for a product. There are two primary types of CAPP systems—variant and generative.

The variant type, which represents the vast majority of such systems currently in use, relies heavily on group technology (GT).

In group technology, a manufacturer classifies parts produced according to various characteristics, e.g., shape, size, material, presence of teeth or holes, and tolerances. In the most elaborate group technology systems, each part may have a 30- to 40-digit code. Group technology makes it easier to systematically exploit similarities in the nature of parts produced and in machining processes to produce them. The theory is that similar parts are manufactured in similar ways. So, for example, a process planner might define a part, using group technology classification techniques, as circular with interior holes, 6 inch diameter, 0.01 inch tolerance, and so forth. Then, using a group technology-based CAPP system, the planner could recall from computer memory the process plan for a part with a similar group technology classification, and edit that plan for the new, but similar, part.

Generative process planning systems, on the other hand, attempt to generate an ideal routing for a part based on information about the part and sophisticated rules about how such parts should be handled, and the capabilities of machines in the plant. The advantage of such systems is that process plans in variant systems may not be optimal. A variant system uses as its foundations the best guesses of an engineer on how to produce certain parts. The variants on this process plan may simply be variations of one engineer's bad judgment.

Though generative CAPP may also depend on group technology principles, it approaches process planning more systematically. The principle behind such systems is that the accumulated expertise of the firm's best process planners is painstakingly recorded and stored in the computer's memory.

According to Arthur D. Little, Inc., benefits of CAPP arc:

Reduced new part introduction cost
Standardized routings
Utilized more optimal processes
Reduced need for process engineers

Lockheed-Georgia developed a generative CAPP system called Genplan to create process plans for aircraft parts (see photo for an example of a process plan developed by Genplan). Engineers assign each part a code based on its geometry, physical properties, aircraft model, and other related information. Planners can then use Genplan to develop the routing for the part, the estimated production times, and the necessary tooling. Lockheed-Georgia officials report that one planner can now do work that previously required four to eight people, and that a planner can be trained in one year instead of three to four.

Plessey implemented a CAPP system from American Channels. Before the installation, there was little control over who did what. Implementation began in early 1985, and by early 1986, all new products were created on the system. On jobs run through the CAPP system, Plessey realized a 40% productivity increase on process plans, and 65% on work study. The CAPP system caused lead time to drop 90% from two weeks to one day.

BIBLIOGRAPHY

Committee on the CAD/CAM Interface, National Research Council (1984). *"Computer Integration of Engineering Design and Production, A National Opportunity."* Washington, D.C., National Academy Press.

Office of Technology Assessment (1984). *Computerized Manufacturing Automation: Employment, Education, and the Workplace* (Washington, DC: U.S. Congress, OTA-CIT-235, April).

Miller, R.K. (1986). *Strategic Planning for Computer Integrated Manufacturing*, SEAI Technical Publications, Madison, GA 1986.

4

Machine Vision

4.1 Solid-State Cameras
4.2 Resolution
4.3 Lighting
4.4 Connectivity Analysis
4.5 Binary Processing and Gray Scale
4.6 Three-Dimensional Vision
4.7 Commercial Machine Vision Systems
4.8 Future Trends
Bibliography

A machine vision system is composed of a light source and camera to capture an image, a computer controller to analyze and process the image, and an output device or devices to relay process control instructions, print reports, or take physical action, such as alerting an operator or correcting the process.

In functional terms, image capture (digitization of the image) by the camera is followed by image preprocessing by the computer controller. The computer then analyzes the stored, preprocessed image by subjecting it to a number of preprogrammed routines. The contrasting features of the image can be enhanced, if necessary, to facilitate image analysis. If the image fails to pass one or more of the tests, action (usually in the form of a specific signal) is initiated by the vision system.

Machine vision can be broken down into the following general categories:

1. Gaging.   Performing precise dimensional measurements.

2. Verification.   Qualitatively ensuring that one or more desired features are present and/or undesired features are absent.

3. Flaw detection.   Finding and discriminating unwanted features of unknown size, location, and shape.

4. Identification.   Determining the identity of an object from symbols, including alphanumeric characters.

5. Recognition.   Determining the identity of an object from observed features of the object.

6. Locating.   Determining the location and orientation of an object.

Three characteristics of machine vision systems of primary importance in industrial tasks are resolution, processing, and speed.

4.1 SOLID-STATE CAMERAS

Silicon detectors called charge-coupled devices (CCDs) were invented at Bell Laboratories in 1969 by Willian Boyle and George Smith. These devices generate an electronic signal proportional to incident light. Silicon is known to absorb photons in the range of 200 to 1100 nm. Charge-coupled devices are self-scanning, with precision permanently etched in its silicon structure, and transmit signals representing the scene being analyzed in periodic discrete "packets" of information easily understood by the interfaced computer.

Operationally, a solid-state image sensor converts incident light to electric charge which is integrated and stored until readout. The integrated charge is directly proportional to the intensity of the light impinging on the sensing elements. Readout is initiated by a periodic start or transfer pulse. The charge information is then sequentially read out at a rate determined by clock pulses applied to the image sensor. The output is a discrete time analog representation of the spatial distribution of light intensity across the array.

Charge-coupled devices are manufactured into matrix, linear and circular arrays. The first two types are of most interest for machine vision systems, producing two-dimensional and one-dimensional images of scenes respectively. A solid-state pixel array will generate a representation or an entire scene or a window of a scene.

Fig. 4.1 Examples of solid-state charge-coupled device pixel arrays. (Courtesy of EG&G Reticon.)

A linear array may be used for objects which are in relative motion to the camera, such as parts moving on a conveyor.

Examples of some solid-state charge-coupled device pixel arrays are shown in Figure 4.1.

4.2   RESOLUTION

The spatial resolution of an imaging system refers to the number of pixels into which the original image is digitized. Most vision systems currently available can resolve images into arrays having spatial resolutions ranging from 128×128 (16,384 pixels per image) to 512×512 (262,144 pixels per image).

Currently, most machine vision systems process a 320×480 matrix. This results in a pixel of 1/320 of the field of view, or 0.3%. In applications requiring greater definition, the object must be scanned with a linear array scan camera, in sections with multiple cameras or through a step and repeat process. A step and repeat scanner can divide the object into sections to be scanned one by one by using X-Y translation. Processing speed, of course, is proportional to both the number of pixels and object positioning time.

There are no known rules limiting resolution. Camera pixel size generally determines resolution for any particular measurement. Charge-coupled device cameras with a 1035×1320 array were introduced in 1986, representing the state-of-the-art (see Figure 4.2). Various algorithms have been developed to achieve resolution of less than one half pixel through subpixel techniques. The ability of vision systems to achieve high resolution was demonstrated at Carnegie-Mellon in an application which measured 4 micron feature size with a resolution of 0.1 micron with a camera system. To achieve high accuracy measurement for large objects, windowing is often necessary.

4.3   LIGHTING

Some of the common illumination sources are tungsten, quartz halogen, quartz iodine, fluorescent, and mercury (or xenon) arc lamps,

Fig. 4.2 These images simulate the difference in resolution between the Videk Megaplus camera (left) with 1.4 million pixels, and a camera having 256 × 300 lines of resolution (right).

as well as various flash lamps, lasers and light emitting diode (LED) sources. The common ways to configure these sources are: front light or spot and back light or spot, as well as collimated back lighting. The line illumination is used with a linear solid-state image sensor while the spot and collimated sources can be used with either the linear or matrix solid-state image sensor. Figure 4.3 illustrates such object-lighting techniques as front lighting, rear lighting, spectral illumination, spectral elimination, beam splitting, split mirror, offset shadowing, and collimated light.

When possible, rear lighting is preferred since it provides greater image contrast. Front lighting must be used, however, where surface features must be extracted. Light intensity must be sufficient to swamp interferences from ambient sources.

Image contrast is also a constraint. The contrast of the object against its background must be greater than the local lighting variation around a feature of interest. Lighting variations are caused by point light sources and interference from ambient light. Features that the system must extract, such as edges or holes, should be dis-

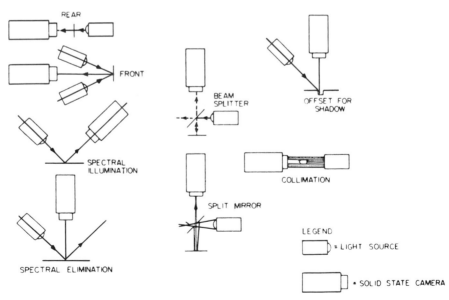

Fig. 4.3   Basic lighting techniques.

tinguished from the local background by 15 to 25% of the overall image intensity range for reliable detection. Thus, for example, using a brightness scale of 1 to 10, a system can distinguish an edge if local background intensity is at level 3 and the edge is illuminated with at least 4.5. Nominal edge-level intensity should be 5.5 or greater.

Structured light is the use of sheets of light and other projective light configurations to directly determine shape and/or range from the observed configuration that the projected line, circle, grid, etc. makes as it intersects the object. An example of structured lighting for machine vision is the system developed by General Motors for their CONSIGHT system.

A slender tungsten bulb and cylindrical lens are used to project a narrow and intense line of light across the belt surface. The line camera is positioned so as to image the target line across the belt. When an object passes into the beam, the light is intercepted before it reaches the belt surface (Figure 4.4). When viewed from above, the line appears deflected from its target wherever a part is passing on the belt. Therefore, wherever the camera sees brightness, it is viewing the unobstructed belt surface; wherever the camera sees darkness, it is viewing the passing part.

Unfortunately, a shadowing effect causes the object to block the light before it actually reaches the imaged line, thus distorting the part image. The solution is to use two (or more) light sources all directed at the same strip across the belt (Figure 4.4). When the first light source is prematurely interrupted, the second will normally not be. By using multiple light sources and by adjusting the angle of incidence appropriately, the problem is essentially eliminated.

4.4 CONNECTIVITY ANALYSIS

Vision technology is largely based on research done at SRI International in the early 1970s, where the technique of connectivity analysis, so-called because it breaks a binary image into its connected components, was developed. The connectivity analysis program builds a description of each blob (a connected component, either

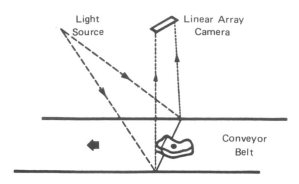

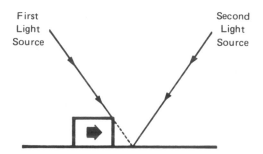

Fig. 4.4 CONSIGHT vision system developed by General Motors, showing basic lighting principle and improvement with two light sources.

an object or a hole) as the image is processed. An array is created to hold information about the blob and its shape. Finally, a number of shape and size feature values characterizing the blob are derived: maximum and minimum values of width and height, area, perimeter height, holes, centroid position, moments of inertia, orientation (from second moments), elongation index, compaction index, and a linked list of coordinates on the perimeter. The connectivity analysis is performed in parallel with run-length encoding, a data compression technique in which an image is scanned faster and only the lengths of "runs" of consecutive pixels with the same color are stored. The processing time is directly proportional to the image area. Today's vision systems utilize this analysis as a basis;

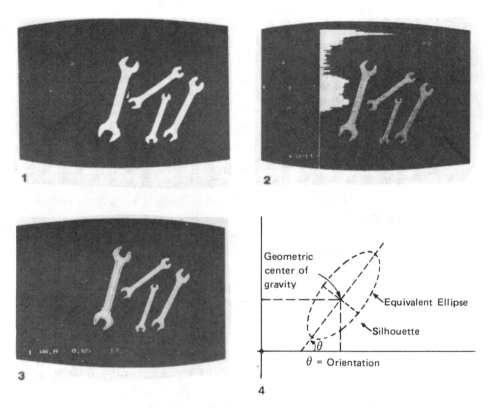

Fig. 4.5 Example showing connectivity analysis.

however, each manufacturer has advanced the basic technique with proprietary algorithms.

Recognition involves several steps (see Figure 4.5). First, the objects in the television camera image are reduced to their silhouettes by setting all the gray-scale intensities in the object's background to black and all the object intensities to white (Miller, 1986). This silhouetting is done as the image is scanned from the TV camera, using a technique called thresholding. The system assumes that objects of interest will contrast sharply with the background. Thus all object intensities will be above or below a certain level, depending on whether the objects are lit from the front or the back. This threshold value is used to determine which intensities should be set to white and which to black during silhouette formation.

The threshold value itself is determined by the operator during system training. As an aid to threshold determination, the vision system displays a histogram of the intensities encountered in a typical scene (Meyer, 1986). If the scene has sufficient contrast, the histogram will have two peaks—a dark one for the background and a bright one for the objects. The threshold value will then be the bottom of the "valley" between the peaks.

The outlines of the silhouettes are traces (Zuech and Miller, 1987). Tracing is done by systematically scanning the image for silhouette edges, starting in the upper left-hand corner and moving line by line to the botton. (Actually, the system scans a compressed version of the image to minimize scanning time.) The compressed image contains the beginning and end locations of continuous runs of the same grey level—the only information needed to determine silhouette edges. This is called run-length encoding. A transition from black to white (left edge) or white to black (right edge) signals the edges of a white silhouette on a black ground. Whenever the computer encounters an edge point, it determines what edge the point belongs to by examining neighboring points on the line above. It then enters the point's location in a list for that edge. If the point is isolated, however, the system assumes a new silhouette and creates a new list.

Because the computer always scans the image from top to bottom, it can never be sure that the edge it encounters through the scan is not connected further down in the image. For example, if the system first encounters the line of a fork, it will treat the lines as separate objects until it reaches the fork's palm. For this reason, edge lists are always provisional until a scan is completed. Lists are consolidated when edges are found to be connected, which happens when the system discovers a point common to two or more edges.

Next the system computes the location and orientation of gravity; its orientation, as the orientation of an ellipse, that has the same area (Rosen and Nizan, 1975), or in terms of some other geometric property.

Finally the system attempts to match the silhouettes to the examples stored in its memory. A close match is considered recognition. The closeness of a match is determined by scoring individual scores to create a total score. By adjusting the weighting factors, it is possible to recognize objects with variable features.

4.5 BINARY PROCESSING AND GRAY SCALE

The vision processing where pixel values are recorded as either "0" or "1" is referred to as binary processing. This technique is a "black and white" type of analysis. It can be used for geometric analysis and edge detection, but cannot be utilized for analysis which requires that surface characteristics be quantified.

For advanced analysis, information may be required to aid complex part recognition or for the analysis of surface characteristics (i.e., texture, shade, pattern, etc.). The gray level is a quantized measurement of image irradiance or brightness. The representation of the image as an array of brightness is obtained from the digitizer. Various vision systems utilize different numbers of gray levels. For analysis by a 16-bit microprocessor, gray-level scales are generally even digital power numbers: 4, 16, 64, or 256.

A comparison of binary and gray-scale processing is presented in Figure 4.6.

4.6 THREE-DIMENSIONAL VISION

A shortcoming of today's machine vision systems is that they analyze only the two-dimensional projected image of a scene. The detail that they see is analogous to that captured in a photograph. Research at MIT, JPL, and other laboratories has been developing two-camera systems which utilize parallax to provide a three-dimensional image. Such systems are now commercially available from a few companies.

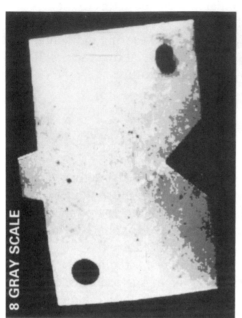

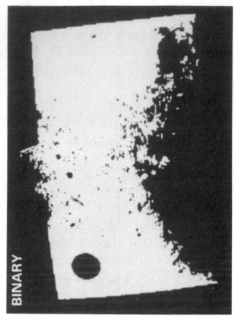

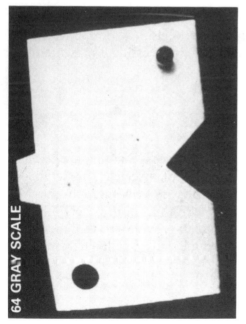

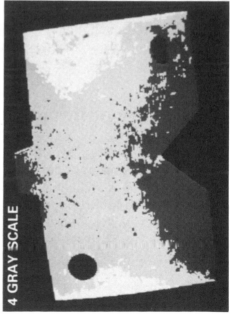

4.7 COMMERCIAL MACHINE VISION SYSTEMS

The image processing/decision making speed today is limited by computer technology. The amount of information associated with complex parts or scenes, number of gray levels, and use of multiple cameras all tend to increase the amount of information which must be processed by the computer. Companies that offer parallel-pipeline image processing architecture (such as Applied Intelligent Systems, Machine Vision International, and Synthetic Vision Systems) claim the highest processing rates, 5 to 40 million instructions per second (MIPS). Some companies have developed co-processor techniques and, where applicable, have the ability to handle up to 40 MIPS. Recently, two international companies (Matra and Mitachi) with links to the semiconductor business have announced the development of special integrated circuit chips that perform some image processing functions at rates up to 10 MIPS. As a point in passing, it is estimated that humans process image data at a rate of 30 to 100 billion instructions per second.

A variety of proprietary algorithms are available from machine vision companies. Most commercial vision software is based on three basic techniques. The first class of algorithms to be used commercially involved connectivity analysis, developed in the 1970s at SRI International. While still in popular use, connectivity analysis has given way to correlation analysis. Normalized correlation is most popular in the industry right now. Cellular processing or mathematical morography, a technique developed originally at Carnegie-Mellon, and perfected at the University of Michigan, is successfully used in at least three commercial systems.

Fig. 4.6 This series of photographs show clearly the advantages of high intensity resolution. With 64 levels of gray, part boundaries are distinguishable from background; and fine features, such as the shadow cast by the post in the lower right-hand corner, are discernible. Using fewer levels of gray renders the scene more subject to the effects of shading (note the deterioration of the lower, slightly shaded portion of the part), and obliterates subtle features. (Courtesy of Analog Devices, Norwood, Massachusetts.)

Commercial vision systems have improved significantly in programmability within the past two years. Still, virtually all applications require a lot of custom software development. This task is sufficiently difficult to require vendor contract support or a systems house for most installations.

For simple inspection tasks, machine vision systems can perform faster than objects can be physically presented to a camera. However, processing speed is a major limitation for complex applications. While some commercial systems can process at a few billion cellular operations per second, and 40 billion per second has been achieved in the laboratory, researchers suggest that minimum array processing systems must be on the order of a TERAflop to be really useful.

4.8 FUTURE TRENDS

The implementation of machine vision is in its infancy. The present installed base of machine vision sites in the United States is estimated at 10,000, dwarfed by the more than 1,000,000 potential applications for this technology.

In June 1986, the Automated Vision Association published *Machine Vision, A Delphi Forecast to 1990.* Among the forecast highlights are:

a. Sales of machine vision products totalled $58 million in 1985 and will reach $457 million in 1990.
b. Systems incorporating machine vision will grow from $188 million in 1985 to more than $2 billion in 1990.
c. While most vision units today (60%) are employed in stand-alone applications, the reverse is expected to be true by 1990 when some 70% of machine vision units are expected to be tied into an integrated cell or system.
d. By 1990, the electronics industry is predicted to be the largest purchaser of machine vision systems, with a 36% share of the market.

e. Gaging functions comprise the largest share of the vision market today, with inspection expected to comprise as large a share by 1990.

f. Depending on application, the cost of a typical basic machine vision unit ranges from between $25,000 and $50,000 at present, with the range expected to fall to $15,000 to $35,000 by 1990.

g. While video will remain the predominant type of image sensor (80% in 1985, down to 64% in 1990), laser scanning is anticipated to increase significantly (from 10% today to 22% in 1990).

h. Only 2% of all vision units are currently equipped to utilize color. By 1990 this figure is expected to reach 15%.

i. While only 5% of present vision units can perform three-dimensional analysis, the comparable number for 1990 is predicted to be 30%.

j. Emphasis in near-future research and development are expected to center upon data processing capabilities, artificial intelligence, camera technology, real-time guidance, three-dimensional techniques, and color analysis.

The major area of effort in vision research is algorithm development. These efforts are generally aimed at achieving higher resolution through subpixel processing, correlation transforms, and gray scale techniques. It has been observed that commercial vision systems all operate on simple concepts that are applied very cleverly. Advanced image processing technology, scene analysis, and artificial intelligence have not yet been applied to any significant extent. These techniques may be expected in future generation systems— increased processing speeds will be necessary to maintain real-time operation.

The achievement of increased processing speeds will be achieved through the continuing evolution of VLSI technology. Some researchers, as well as the U.S. Department of Defense DARPA program, are looking at parallel-processing architectures for achieving dramatic increases in processing speeds. Optical pattern recognition is showing promise, and it is likely that some future commercial systems will employ optical rather than electronic processing. Grumman, for example, has developed an experimental vision system which directs robotic bin picking by use of optical correlation.

BIBLIOGRAPHY

Meyer, J.D. (1986). *Machine Vision Systems*: *A Summary and Forecast*, Tech Tran Consultants, Lake Geneva, WI.

Miller, R.K. (1986). *Machine Vision for Robotics and Automated Inspection*, 3 volumes. SEAI Technical Publications, Madison, GA.

Rosen, C.A. and Nizan, D. (1975). "Some Developments in Programmable Automation," *Proc. IEEE Intercom 75*, 09 April.

Ward, M.R., Rossol, L., and Holland, S.W. (1979). "Consight: An Adaptive Robot with Vision," *Robotics Today*, 26-32, Summer.

Zuech, N., and R.K. Miller. (1987). *Machine Vision*, Prentice-Hall, Englewood, NJ.

5

Sensors for Industrial Inspection

5.1   Introduction
5.2   Ultrasonic Sensors for Automated Inspection
5.3   Automatic Inspection for Packaged Products

5.1 INTRODUCTION

An extremely wide range of sensors are used for automated quality assurance. The largest class of automated quality assurance (AQA) sensors are classified as machine vision systems. Machine vision is presented in Chapter 6 of this book. Other sensor technologies and their application to specific AQA products are discussed in this chapter.

Some applications of industrial sensors include hardness, case depth, conductivity, metal sorting, coating/plating thickness, machinability, cracks/seams, weld/bond strength, porosity/inclusions/chevrons, wall thickness, dimensions/measuring, level of fill, process

Table 5.1   Inspection Technologies.

METHOD	MEASURES OR DEFECTS	APPLICATIONS	ADVANTAGES	LIMITATIONS
PENETRANTS (Dye or fluorescent)	Defects open to surface of parts; cracks, porosity, seams, laps, etc. Through-wall leaks	All parts with non-absorbing surfaces (forgings, weldments, castings, etc).	Portable - Low cost Indications may be further examined visually Results easily interpreted	Surface films, such as coatings, scale, & smeared metal may prevent detection of defects
RADIOGRAPHY (thermal neutrons from reactor, accelerator, or Californium-252	Hydrogen contamination of titanium or zirconium alloys Defective or improperly loaded pyrotechnic devices Improper assembly of metal, nonmetal parts Corrosion products	Pyrotechnic devices Metallic, nonmetallic assemblies Biological specimens Nuclear reactor fuel elements & control rods Adhesive bonded structures	High neutron absorption by hydrogen, boron, lithium, cadmium, uranium, plutonium Low neutron absorption by most metals Complement to X-ray or gamma-ray radiography	Very costly equipment Nuclear reactor or accelerator required Trained physicists required Radiation hazard-Nonportable Indium or gadolinium screens required
RADIOGRAPHY (gamma rays) Cobalt-60 Iridium-192	Internal defects & variations; porosity, inclusions, cracks, lack of fusion, geometry variations, corrosion thinning Denisty variations Thickness, gap & position	Usually where X-ray machines are not suitable because source cannot be placed in part with small openings and/or power source not available Panoramic imaging	Low initial cost Permanent records; film Small sources can be placed in parts with small openings Portable - Low contrast	One energy level per source Source decay Radiation hazard Trained operators needed Lower image resolution Cost related to source size
RADIOGRAPHY (X-rays - film)	Internal defects & variations; corrosion thinning Density variations Thickness, gap & position	Castings - Weldments Electrical assemblies Small, thin, complex wrought products Nonmetallics Solid propellant rocket motors	Permanent records; film Adjustable energy levels (5kv-25mev) High sensitivity to density changes No couplant required	High initial costs Orientation of linear defects in part may not be favorable Radiation hazard Depth of defect not indicated
RADIOMETRY (X-ray, gamma-ray, beta-ray) (Transmission or backscatter)	Wall & Plating thickness Variations in density or composition Fill level in cans or containers Inclusions or voids	Sheet, plate, foil, strip, tubing Nuclear reactor fuel rods Cans or containers Plated parts Composites	Fast - Fully automatic Extremely accurate In-line process control Portable	Radiation hazard Beta-ray useful for ultrathin coatings only Source decay Reference standards required

Method	Measures	Materials	Advantages	Limitations
SONIC (Less than 0.1 MHz)	Debonded areas or delaminations in metal or nonmetal composites or laminates; Cohesive bond strength under controlled conditions; Crushed or fractured core	Metal or nonmetal composite or laminates brazed or adhesive-bonded; Plywood; Rocket motor nozzles; Honeycomb	Portable - Easy to operate; Locates far-side debonded areas; May be automated; Access to only one surface required	Surface geometry influences test results; Reference standards required; Adhesive or core thickness variations influence results
THERMAL (thermochronic paint, liquid crystals)	Lack of bond - Hot spots; Heat transfer - Isotherms; Temperature ranges; Blockage in coolant passages	Brazed joints; Adhesive-bonded joints; Metallic platings or coatings; Electrical assemblies; Temperature monitoring	Very low initial cost; Can be readily applied to surfaces which may be difficult to inspect by other methods; No special operator skills	Thin-walled surfaces only; Critical time-temperature relationship; Image retentivity effected by humidity
THERMO-ELECTRIC PROBE	Thermoelectric potential; Coating thickness; Phsical properties; Thompson effect; P-N junctions in semiconductors	Metal sorting; Ceramic coating thickness on metals; Semiconductors	Portable; Simple to operate; Access to only one surface required	Hot probe; Difficult to automate; Reference standards required; Surface contaminants; Conductive coatings
TOMOGRAPHY	Boundaries; Surface reconstruction; Crack size, location & orientation	Metals research; Medicine	Pinpoint defect location; Image display is computer controlled	Very expensive; Need highly trained operator
ULTRASONIC (0.1-25 MHz)	Internal defects and variations; cracks, lack of fusion, porosity, inclusions, deaminations, lack of bond-texturing; Thickness or velocity	Wrought metals; Welds - Brazed joints; Adhesive-bonded joints; Nonmetallics; In-service parts	Most sensitive to cracks; Test results known immediately; Automating & permanent record capability; High penetration capability	Couplant required; Small, thin, complex parts may be difficult to check; Reference standards required; Special probes
ULTRASONIC (critical angle reflectivity)	Elastic properties, acoustic attenuation in solids; Near-surface metallic property gradients, e.g. carburization in steel; Metallic rain structure & size	Metals; Nonmetals	Access to only one surface required; Permanent record; Quantitative; No physical contact of sample required; Sample preparation minimal	Test parts must be immersed; Geometry limitations: test part must have a flat, smooth area; Goniometer device required; Skilled technician required

Table 5.1 (Continued)

METHOD	MEASURES OR DETECTS	APPLICATIONS	ADVANTAGES	LIMITATIONS
FLUOROSCOPY (Cine-fluorography) (Kine-fluorography)	Level of fill in containers Foreign objects Internal components Density variations Voids, thickness Spacing or position	Particles in liquid flow Presence of cavitation Operation of valves & switches Burning in small solid-propellant rocket motors	High-brightness images Real-time viewing Image magnification Permanent record Moving subject can be observed	Costly equipment Geometric unsharpness Thick specimens Speed of event to be studied Viewing area
HOLOGRAPHY (Acoustical-liquid surface levitation)	Lack of bond Delaminations Voids - Porosity Resin-rich or resin-starved areas Inclusions Density variations	Metals - Plastics Composites - Laminates Honeycomb structures Ceramics Biological specimens	No hologram film development required Real-time imaging provided Liquid-surface responds rapidly to ultrasonic energy	Through-transmission techniques only Object & reference beams must superimpose on special liquid surface Immersion test only Laser required
HOLOGRAPHY (interferometry)	Strain - Cracks Plastic deformation Debonded areas Voids & inclusions Vibration	Bonded & composite structures Automotive or aircraft tires Three-dimensional imaging	Surface of test object can be uneven No special surface preparations or coatings required No physical contact with test specimen	Vibrationfree environment is required Heavy base to dampen vibrations Difficult to identify type of flaw detected
INFRARED (radiometers)	Lack of bond Hot spots Heat transfer Isotherms Temperature ranges	Brazed joints Adhesive-bonded joints Metallic platings or coatings; debonded areas or thickness Electrical assemblies	Sensitive to 1.5°F temperature variation Permanent record or thermal picture Quantitative	Emissivity Liquid-nitrogen-cooled detector Critical time-temperature relationship

LEAK TESTING	Leaks
	Helium - Ammonia
	Smoke - Water
	Air bubbles
	Radioactive gas
	Halogens
		Joints:
		  Welded - Brazed
		  Adhesive-bonded
		  Sealed assemblies
		Pressure or vacuum chambers
		Fuel or gas tanks
			High sensitivity to extremely small, tight separations not detectable by other NDT methods
			Sensitivity related to method selected
				Accessibility to both surfaces of part required
				Smeared metal or contaminants may prevent detection
				Cost related to sensitivity

Method				
MAGNETIC FIELD	Cracks - Wall thickness Hardness - Coercive force Magnetic anisotropy Magnetic field Nonmagnetic coating thickness on steel	Ferromagnetic materials Ship degaussing Liquid level control Treasure hunting Wall thickness of non-metallic materials	Measurement of magnetic material properties May be automated Easily detects magnetic objects in nonmagnetic material Portable	Permeability Reference standards required Edge-effect Probe lift-off
MAGNETIC PARTICLE	Surface & slightly subsurface defects; cracks, seams, porosity, inclusions Permeability variations Extremely sensitive for locating small tight cracks	Ferromagnetic materials; bar, forgings, weldments, extrusions, etc.	Advantage over penetrant in that it indicates subsurface defects, particularly inclusions Relatively fast & low cost May be portable	Alignment of magnetic field is critical Demagnetization of parts required after tests Parts must be cleaned before & after inspection
MAGNETIC PERTURBATION	Cracks - Crack depth Broken strands in steel cables Permeability effects Nonmetallic inclusions	Ferromagnetic metals Broken steel cables in reinforced concrete	May be automated Easily detects magnetic objects in nonmagnetic materials Detects subsurface defects	Requires reference standard Need trained operator Part geometry Expensive equipment
MICROWAVE (300 MHz-300 GHz)	Cracks, holes, debonded areas, etc. in nonmetallic parts Changes in composition, degree of cure, moisture content Thickness measurement Dielectric constant	Reinforced plastics Chemical products Ceramics - Resins Rubber - Liquids Polyurethane foam - Radomes	Between radio waves & infrared in the electro-magnetic spectrum Portable Contact with part surface not normally required	Will not penetrate metals Reference standards required Horn to part spacing critical Part geometry Wave interference - Vibration
MOSSBAUER EFFECT	Nuclear magnetic resonance in materials, most common being iron-57 Polarization of magnetic domains in steel	Detect & identify iron in specimen or sample Detect iron films on stainless steel Measure retained austenite	Provide unique information about the surroundings of the iron-57 nuclei	Radiation hazard Trained engineers or physicists required Nonportable
NEUTRON ACTIVATION ANALYSIS (Reactor, accelerator, or radio-isotope)	Radiation emission resulting from neutron activation Oxygen in steel Nitrogen in food products Silicon in metals & ores	Metallurgical - Prospecting Well logging Oceanography On-line process control of liquid or solid materials	Automatic systems - Fast Accurate (ppm range) No contact with sample Sample preparation minimal	Radiation hazard Fast decay time Reference standard required Sensitivity varies with irradiation time

Table 5.1 (*Continued*)

METHOD	MEASURES OR DETECTS	APPLICATIONS	ADVANTAGES	LIMITATIONS
ACOUSTIC EMISSION	Crack initiation and growth rate Internal cracking in welds during cooling Boiling or cavitation Friction or wear Plastic deformation Phase transformations	Pressure vessels Stressed structures Turbine or gear boxes Fracture mechanics research Weldments Sonic signature analysis	Remote and continuous surveillance Permanent record Dynamic (rather than static) detection of cracks Portable Triangulation techniques to locate flaws	Transducers must be placed on part surface Highly ductile materials yield low amplitude emissions Part must be stressed or operating Test system noise needs to be filtered out
ACOUSTIC-IMPACT (tapping)	Debonded areas or delaminations in metal or nonmetal composites or laminates Cracks in turbine wheels or turbine blades Loose rivets or fasteners Crushed core	Brazed or adhesive-bonded structures Bolted or riveted assemblies Turbine blades Turbine wheels Composite structures Honeycomb assemblies	Portable Easy to operate May be automated Permanent record of positive meter readout No couplant required	Part geometry and mass influences test results Impactor and probe must be repositioned to fit geometry of part Reference standards required Pulser impact rate is critical for repeatability
BARKHAUSEN NOISE ANALYSIS	Residual stresses in ferromagnetic steels	Jet engine components such as compressor blades, discs diffuser cases	Nondestructive stress analysis Permanent record Fully automatic	Expensive Requires reference standard Need trained operator Not yet a production tool
EDDY CURRENT (100 Hz to 10 kHz)	Subsurface cracks around fastener holes in aircraft structure	Aluminum and titanium structure	Detect subsurface cracks not detectable by radiography	Part geometry Will not detect short cracks
EDDY CURRENT (10 kHz to 6 MHz)	Surface and subsurface cracks and seams Alloy content Heat treatment variations Wall and coating thickness Crack depth Conductivity Permeability	Tubing/Wire Ball bearings "Spot checks" on all types of surfaces Proximity gage Metal detector Metal sorting Measure conductivity in % IACS	No special operator skills High speed, low cost Automation possible for symmetrical parts Permanent record capability for symmetrical parts No couplant or probe contact required	Conductive materials Shallow depth of penetration (thin walls only) Masked or false indications caused by sensitivity to variations, such as part geometry, lift-off Reference standards required Permeability variations

Method	Applications	Materials	Advantages	Limitations
EDDY-SONIC	Debonded areas in metal-core or metal-faced honeycomb structures Delaminations in metal laminates or composites Crushed core	Metal-core honeycomb Metal-faced honeycomb Conductive laminates such as boron or graphite fiber composites Bonded metal panels	Portable Simple to operate No couplant required May be automated	Speciman or part must contain conductive materials to establish eddy-current field Reference standards required Part geometry
ELECTRIC CURRENT (direct current conduction method)	Cracks Crack depth Resistivity Wall thickness Corrosion-induced wall-thinning	Metallic materials Electrically conductive materials Train rails Nuclear fuel elements Bars, plates, other shapes	Access to only one surface required Battery or dc source Portable	Edge effect Surface contamination Good surface contact required Difficult to automate Electrode spacing Reference standards required
ELECTRIFIED PARTICLE	Surface defects in non-conducting material Through-to-metal pinholes on metal-backed material Tension, compression, cyclic cracks Brittle-coating stress cracks	Glass Porcelain enamel Nonhomogeneous materials such as plastic or asphalt coatings Glass-to-metal seals	Portable Useful on materials not practical for penetrant inspection	Poor resolution on thin coatings False indications from moisture streaks or lint Atmospheric conditions High voltage discharge
EXO-ELECTRON EMISSION	Fatigue in metals	Metals	Access to only one surface required Permanent record Quantitative	No surface films or contamination Geometry limitations Skilled technician required
FILTERED PARTICLE	Cracks Porosity Differential absorption	Porous materials such as clay, carbon, powdered metals, concrete Grinding wheels High-tension insulators Sanitary ware	Colored or fluorescent particles Leaves no residue after baking part over 400°F Quickly & easily applied Portable	Size & shape of particles must be selected before use Penetrating power of suspension medium is critical Particle concentration must be controlled Skin irritation

Source: Courtesy of Inspectech, Holly, Michigan.

control, part recognition, surface finish, leaks, thread quality, stress, temperature, and vibration.

Inspection sensors are used for nondestructive testing, dimensional gauging, instrumentation, and automated systems to eliminate or reduce human error in materials and manufacturing processes. There are 32 specific inspection technologies that may be identified for industrial applications. Table 5.1 presents a comparison of these technologies. The chart was provided by George M. Nygaard, President, Inspectech (Holly, MI).

The following sections examine two specific areas of sensor technology as they apply to automated quality assurance. First, applications of ultrasonic sensors are reviewed. This class of sensors ranks second to machine vision in usage for computer-integrated manufacturing inspection systems. Next, a variety of sensors used for the automatic inspection of packaged products are presented. This type of inspection is essential in food, pharmaceutical, and other industries.

5.2 ULTRASONIC SENSORS FOR AUTOMATED INSPECTION

Ultrasonics provide a convenient method for the inspection of many types of products. Essentially, inspection systems based on ultrasonics ascertain geometric information about a product or object in one, two, or three dimensions. The most basic type of system is the one-dimensional ultrasonic distance measuring system, used for high-accuracy, noncontact gauging. An example system, commercially available from Ultrasonic Arrays, Inc. (Woodinville, WA), is shown in Figure 5.1. The system consists of a single transducer and a control unit. With this system, measurements may be performed on objects with a distance of 0.2 to 24 inches, with an accuracy of 0.001 inches. The use of two transducers allows the same system to function as a thickness measuring system. To obtain a more complex two-dimensional gauging, an array of multiple transducers may be utilized, the object may be moved in a precision path in the sensor field, or the sensors may be moved to multiple positions. Applications for this type of sensor include:

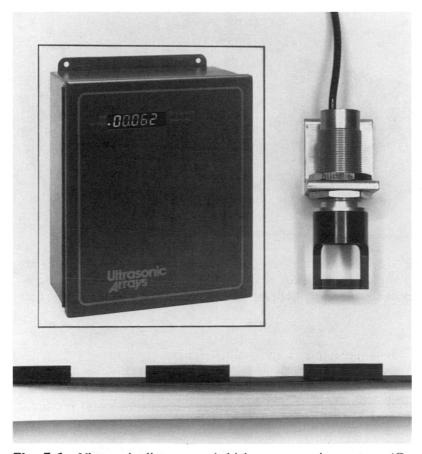

Fig. 5.1 Ultrasonic distance and thickness measuring system. (Courtesy of Ultrasonic Arrays, Woodinville, Washington.)

1. Noncontact measurement of metallic, nonmetallic and liquid substances. Soft substances such as cloth, paper or foam do not affect the accuracy of the measurement.
2. Gauging and inspecting stamped or machined parts.
3. Surface flatness or straightness measurements.
4. Surface perpendicularity, such as whether or not a bottle cap is cross-threaded.
5. Thickness measurements of all types.

Fig. 5.2 Ultrasonic inspection system, with nine-microphone sensor head and controller. Each controller can process inputs from up to six heads simultaneously. (Courtesy of Cochlea Corporation, San Jose, California.)

6. Pipe diameter and wall thickness.
7. Double sheet detection.
8. Piece counting and sorting.
9. Measuring and sizing lumber and other wood products.
10. Solid and liquid level measurements.

An ultrasonic inspection system for three-dimensional measurements is shown in Figure 5.2. This system, available from Cochlea Corporation (San Jose, CA), senses the environment in three dimensions by using an array of ultrasonic transducers.

Some specific applications of ultrasonics to automated inspection follow.

Weld Inspection with Ultrasonics

A mechanized welding head and a computer-controlled ultrasonic search head comprise a new weld inspection system developed at

EG&G Idaho's Materials Science Div. (Idaho Falls, ID) under contract to the U.S. Department of Energy's Idaho National Engineering Laboratory. The ultrasonic head trails behind the welding head, sending the receiving ultrasonic waves which signal weld defects. Immediate inspection allows repair of the problem before additional weld material is applied.

The Concurrent National Department of Energy Welding System, now undergoing laboratory studies, has potential application in shipbuilding and in construction of energy production plants, spacecraft, offshore oil rigs, and pipelines.

Forging Inspection by Ultrasonics

An ultrasonic immersion system has been created to comply with the positioning demands of near net shape jet engine forgings.

The Midus controller was engineered for ease of use and features programs to teach contour-following moves. The longitudinal beam contour is automatically converted to circumferential, axial or radial shear wave inspection upon operator entry of the shear wave angle.

Ultrasonic inspection scan plans for complete part coverage are developed by an operator by linking scan segments. A complete scan can be stored in PROM cassettes. Setup changes can be entered to compensate for part shape or size variations and programs may be interrupted to evaluate indications.

The system was developed by Automation Industries Division of Sperry (Chatsworth, CA).

Ultrasonic Panel Inspection

Six tandem pairs of through-transmission water squirters simultaneously inspect large composite panels and bonded structures at the rate of 90 square feet (8 square meters) per hour in a unique ultrasonic test system installed at Rockwell International (Tulsa, OK).

Designed and manufactured by the Sperry Products Division of Automation Industries (Chatsworth, CA) for the B1 Bomber program, the high-speed production testing system inspects panels as large as 18 feet (5 meters) long by 10 feet (3 meters) wide.

By using distributed processors for each axis, a Sperry Products MIDUS microcomputer provides high-speed control of all axes of motion. Ultrasonic technicians can easily set up the controller through a step-by-step menu to access up to 39 discrete scan plans at the touch of a button. Additionally, chaining any number of those individual scan plans together will provide a single continuous sequence for a large complex part.

Data acquisition is performed by a scanning of a test part. The system is shown in Figure 5.3.

Fig. 5.3   High-speed ultrasonic production testing system inspects large composite panels and bonded structures at Rockwell International. (Courtesy of Sperry Automation Division, Chatsworth, California.)

Ultrasonic Leak Detection

Combining two previously separate manual operations, an innovative ultrasonic leak detection system has boosted hourly production to 700 pans per hour on one "big three" automaker's oil pan production line (see Figure 5.4).

Previously, the line relied upon one worker to manually insert and torque oil pan drain plugs and another to locate the pan drain plugs and another to locate the pan prior to initiating a test cycle. With the new AcousTechs system, designed and manufactured by Oakland Engineering (Pontiac, MI), a single worker inserts a drain

Fig. 5.4 Machine for leak testing of automotive oil pans measures air escaping from sealed cavities. (Courtesy of Oakland Engineering, Pontiac, Michigan.)

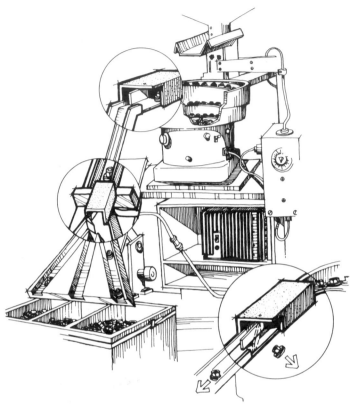

(a)

Fig. 5.5 (a) Ultrasonic inspection system for small parts. (b) Some typical parts which may be inspected. (Courtesy of Cochlea Corporation, San Jose, California.)

plug and places the pan onto a self-centering positioning guide at the machine load station.

Once photo cells detect part presence, the automatic cycle begins at a nutrunner station where drain plugs are torqued. The part is then cycled into the leak detection station, positioned onto gasketed tooling, and clamped, forming a sealed cavity. In less than five seconds, a fast-fill air system pressurizes the cavity to approximately 20 psi (138 kPa) and accepts or rejects the part. In the case of a reject, the system further determines if the leak is from the drain plug assembly or from draw spits or cracks in the part's skin.

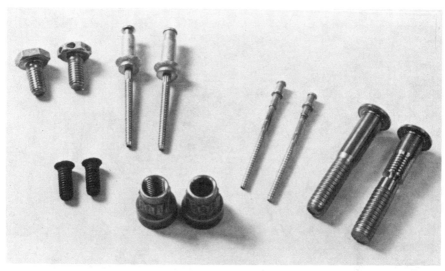

(b)

Fig. 5.4 (*Continued*)

Oakland AcousTechs technology is based upon the fact that pressurized air escaping through a leak generates considerable ultrasonic energy. Detecting this noise in the 30 to 40 kHz range—beyond the range of human hearing—electronic equipment converts the ultrasonic energy into usable signals.

Ultrasonic Inspection of Small Parts

The Sonovision system, from Cochlea Corporation, may be used to inspect a wide range of small- and medium-sized parts (Figure 5.5). Acoustic sensing is found to usually be more precise than optoelectronics for measuring flatness, depth of holes, internal dimensions of bores, and deviations from standards. The figure shows two ways the system could be used in feeding, inspecting, and sorting. At top, one sensor head coarsely sorts and orients parts as they leave the feeder, deflecting misoriented or bad parts back into the bowl (as shown in the insert at lower right). At left center, a second sensor head inspects the parts and allows only good parts to pass

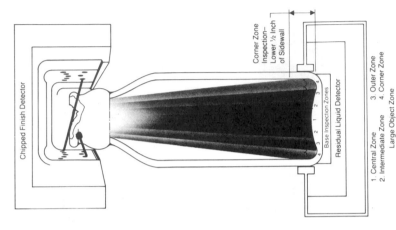

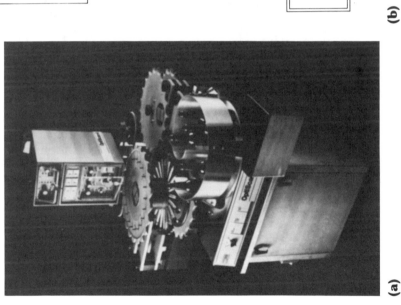

(a)

(b)

Fig. 5.6 (a) OptiScan 4 system for empty bottle inspection. (b) Optical detection method. (Courtesy of BW Electronics. Clearwater, Florida.)

through. In qualification applications, parts usually do not need to be oriented.

5.3 AUTOMATIC INSPECTION FOR PACKAGED PRODUCTS

To meet demands for quality assurance in the food industry, a wide range of automated inspection systems have been developed.

Empty bottles are inspected for geometry, particles, plugged neck, residual liquid, chipped heals, and chipped sealing rings using the OptiScan 4 system (see Figure 5.6), available from BW Electronics (Clearwater, FL). An installation in a major bottling plant in the midwestern United States resulted in the rejection of more than three times the number of defective and contaminated bottles. In addition to increased product quality, the system resulted in filler shutdowns die to bottle failures by a factor of four. Similar systems are being used to check containers ranging in size from large coffee and pcanut butter jars to small infant formula containers.

Fill level inspection is performed automatically by the Filtec systems shown in Figure 5.7, available from Industrial Dynamics Company, Ltd. (Torrance, CA). Sensors and reject systems are selected to match the product and container. Gamma radiation is applicable for most products in metal, paperboard, foil lined, glass, and opaque plastic containers as well as transparent containers. A sealed Americium-241 radioisotope is approved by the Federal Drug Administration for food use. For liquid and powered products in transparent to nearly opaque containers, infrared radiation may be used. Application of infrared is limited to nonfoaming products. Visible light may be used for most transparent or translucent liquids in transparent or translucent glass or plastic containers.

Another type of inspection is for foreign particles. Eisai Co., Ltd., a major pharmaceutical manufacturer in Japan has developed the automated system shown in Figure 5.8 for this application. The inspection system is designed to replace naked-eye inspection for detection of foreign matter in injectables, from small ampules and vials to large infusion bottles. Eisai's Automatic Inspection Machine

Fig. 5.7 Filtec system for automatic inspection of fill level in bottles (top) and cans (bottom). (Courtesy of Dynam, Co., Ltd., Torrance, California.)

66

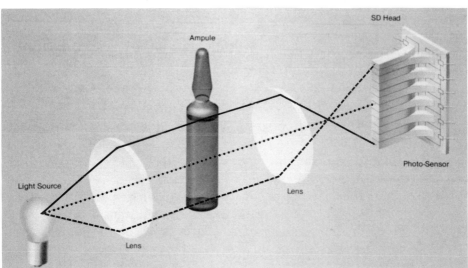

Fig. 5.8 (Top) Eisai Automatic Inspection Machine. (Bottom) The detection system. (Courtesy of Eisai Co., Ltd., Torrance, California.)

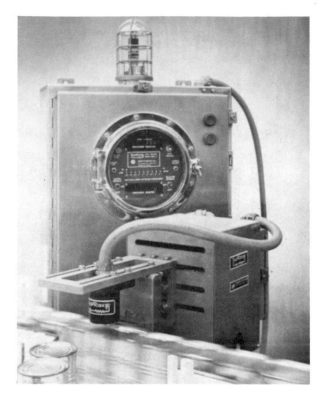

Fig. 5.9 Taptone system for automatic inspection of vacuum in filled containers. (Courtesy of Benthos, Inc., North Falmouth, Massachusetts.)

employs as its detection system the light transmission system illustrated above. This system perceives light variation caused by foreign particulates while they are rotating together with the solution. It was developed on the basis of a most important concept in foreign particulate inspection; namely, to base acceptance or rejection on the size of particulate, not on any other factor such as the reflective strength of the particulate. Furthermore, the light transmission system is free from influence by stains, flaws, labels on the container surface, and so on.

Inspection for vacuum or pressure in packaged food containers (cans, jars, or flexible packaging materials) is desirable for a

wide range of food products. This type of inspection may be performed automatically using an inspection system such as shown in
Figure 5.9. This "Automatic Line Taptone" system, available from
Benthos, Inc. (North Falmouth, MA), detects and reacts to low
vacuum or other container problems before containers are cased. In
an ideal container, a 5% change in vacuum can be detected. Speeds
to 1800 per minute can be achieved. Typical applications include
fruit juices, milk products, and soups. When a defective container
is found, rejection is accomplished by a ram, and a light flashes to
signal that a reject has taken place.

6

Robotic Inspection

6.1   Servo Robots
6.2   Fundamentals of Robot Control and Programming
6.3   Types of Robots
6.4   Reliability
6.5   Intelligent Robots
6.6   Robot Vision
6.7   Robotic Testing and Inspection
6.8   Automobile Body Gauging
6.9   TCM Board Tester
6.10   Valve Testing
6.11   X-Ray Testing
6.12   Detector Alarm Check
6.13   Electrical Contact Testing

The Robotic Industries Association defines an industrial robot as: "A programmable multifunctional manipulator designed to move materials, parts, tools, or specialized devices through variable programmed motions for the performance of a variety of tasks."

Over 20,000 industrial robots are now in use in the United States. There are several reasons for their use and acceptance:

1. Reduced labor costs.
2. Increased output rate.
3. Elimination of dangerous or undesirable jobs.
4. Improved product quality.
5. Increased manufacturing flexibility.
6. Reduced material waste.
7. Easier compliance with OSHA regulations.
8. Reduced labor turnover.
9. Lower capital cost.

6.1 SERVO ROBOTS

The industrial robot consists of two primary elements: (1) the manipulator arm, which moves through the power of several dc electric servo motors, hydraulic servo-valves or other devices, and (2) the controller, which directs motion of the arm by directing power sent to the servo motor or other prime mover. The controller is linked to the robot movement by feedback devices (encoders or resolvers) located in the arm. By comparing these signals related to the robot's actual motion with the programmed instructions, the controller can maintain continuous, or point-to-point, servo-control over the robot motion. In more advanced systems, adaptive control of the robot can be achieved by utilizing artificial sensory input, such as vision or tactile, to allow the robot to modify its motion as appropriate for changing conditions around it.

6.2 FUNDAMENTALS OF ROBOT CONTROL
AND PROGRAMMING

Control of robot arms ranges from highly repeatable open loop devices, to servo-controllers that utilize external sensors in the control of robot actions. Open-loop devices may be as simple as a step sequencer and positionable mechanical stops for a pick and place robot, to more complex devices using stepper motors to reproduce a desired motion. Servo-devices can use internal sensors, such as joint position sensors, or external information sensors such as force, proximity devices, and even vision (operating under computer control). Servos operating only on internal sensors require very careful positioning of the workpiece (which can require expensive fixturing or feedback devices). Servos utilizing data from sensing the external environment, require more complex processing of the signal, but yield much more flexible systems. The controller, in addition to controlling the manipulator motion, often serves as an interface to the outside world—coordinating the robot's motions with machines and assembly lines and turning on and off machines being operated.

Programmable robots are servo controller robots of two basic types: "point-to-point" and "continuous path." Point-to-point robots are directed by a programmable controller that memorizes a sequence of arm and end-effector positions. Hundreds of points may be memorized. The robot moves in a series of steps from one memorized point (set of point positions) to another under servo control, using internal joint sensors for feedback. Because of the servos, trajectory control between the memorized points is possible and relatively smooth motions can be achieved.

Continuous path robots do not depend on a series of intermediate points to generate a trajectory, but duplicate during the playback process the continuous motions recorded during the teaching process. Thus, these robots are used for painting, arc welding, and other processes requiring smooth continuous motions.

Computer-controlled robots are capable of being programmed "off-line" using a high level programming language and do not have to rely on being physically taught.

Sensory robots are computerized robots that interface to the outside world through external sensing such as sight or touch. These "intelligent robots" are capable of adapting to a variety of conditions by changing their goals or preprogrammed decision points.

The most common method for programming a robot is the lead-through method, in which the operator leads the robot through the desired positions and locations by means of a remote teach box.

These points are recorded and used to generate the robot trajectory during operation.

The emphasis in programming research today is on software programming of computer-controlled robots. Work on sensor-controlled manipulation is extending the scope of programmability. Interacting with the robot by means of software provides more flexibility than the other programming methods and allows for conditional actions or flexible adaptations. Various high-level robot programming languages such as VAL (Unimation) and AML (IBM), are now beginning to become available to aid in the software generation.

6.3 TYPES OF ROBOTS

Robot coordinate systems may be of four types (Figure 6.1):

1. Cylindrical.   These robots are basically a horizontal arm with radial traverse mounted on a vertical column with rotation traverse. The work envelope is a portion of a cylinder.

2. Spherical.   These robots have an arm with a radial traverse mounted in a base which pivots and rotates. Their configuration is similar to the turret of a tank. Their work envelope is a portion of a sphere.

3. Jointed-Spherical.   These robots resemble the human arm, with an "elbow" joint and "shoulder" joint. The arm is mounted on a base which provides rotary motion. Their work envelope resembles a distorted sphere. Most new robots being designed today are of this type.

4. Rectangular.   These robots are gantry mounted robots which offer an X-Y-Z traverse. Their work envelope is a rectangle. Robots of these types have appeared on the market only recently and are well suited for assembly and various other localized tasks.

Fig. 6.1   Four types of robot coordinate systems: (A) cylindrical, (B) spherical, (C) jointed-spherical, and (D) rectangular.

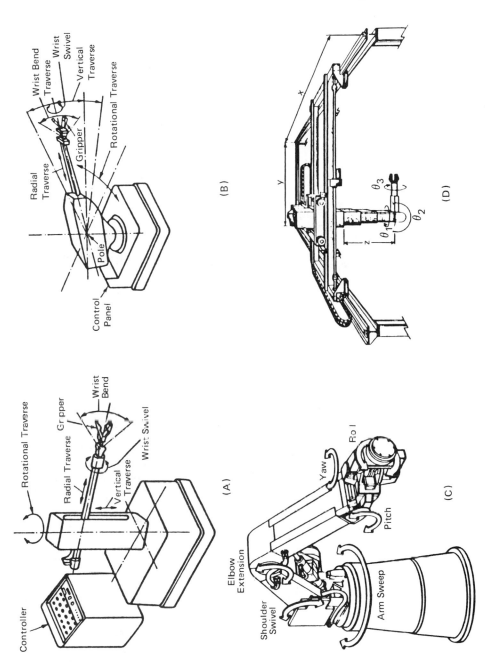

6.4 RELIABILITY

Reliability is one of the most important aspects of a robot installation, since a robot failure may mean stopping an entire production line, and leaving expensive labor and costly equipment idle. Three descriptors are typically used to quantify reliability:

MRBF: mean time between failure
MTTR: mean time to repair
Availability of uptime:

$$\frac{MTBF - MTTR}{MTBF} \times 100\%$$

An uptime of 98% and higher has been achieved by some robot manufacturers.

6.5 INTELLIGENT ROBOTS

Real world industrial environments pose some challenges that the basic industrial robot is unable to cope with, such as picking parts from a bin, processing out-of-place parts, parts out of dimensional tolerance, etc. Therefore, it is desirable for the robot to interface with the outside world through external sensing such as sight or touch. These "intelligent robots" are capable of adapting to a variety of conditions by changing their actions based on relating the sensed information to their goals or programmed decision points. Vision and tactile sensing are the two primary methods used to make robots intelligent.

6.6 ROBOT VISION

Robot vision systems transform information on the orientation of objects into robot coordinates. This enables the robot to identify the points of an object that it can grasp. The robot can then pick up the objects and move them as required. The machine vision systems utilized are the same as for inspection (see Chapter 4)—algorithms simply adapt the measured data for robot control rather than inspection.

Virtually every major manufacturer of industrial robots offer vision systems as an option for their robots. In addition, several other companies produce vision systems which may be used by robots or for inspection tasks.

6.7 ROBOTIC TESTING AND INSPECTION

Most industrial testing operations involve some type of manipulation of either an instrument or the object being tested. Industrial robots have been frequently used to perform the material handling associated with this process.

Since the target function of inspection is to detect manufacturing imperfections, it is imperative that the inspection process itself be frcc of error. Industrial robots have an advantage over human operators in the performance of inspection and testing in that, once programmed, robots will perform a task in precisely the same manner every time.

Robots may be utilized for testing in two ways: The robot handling a test instrument, or the robot handling a component and moving it into appropriate position for instrumentation measurement.

The robotic inspection work cell can range from only a robot and a test instrument to a complete inspection line. Figures 6.2 and 6.3 show two simple work cells, where a robot picks up a product from a production line, tests it, and returns it to the line. In Figure 6.2, a robot at Westinghouse weighs metal oxide blocks used in lightning arresters. Figure 6.3 shows a robot at Saab transferring a metal part for inspection by a machine vision system to insure that all the holes have been drilled and that all the parts have been installed. A more complex installation involving a complete inspection line at Hitachi is shown in Figure 6.4. The rail-mounted robot moves two types of workpieces through six test stations and sorts the "good" and "no good" parts at an unloading station.

Figure 6.2 Robotic transfer of workpiece to instrument. (Courtesy of Westinghouse.)

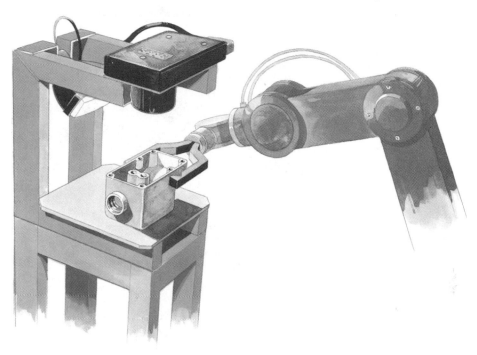

Fig. 6.3 Robot transferring component to machine vision station to verify that all the holes have been drilled and have correct dimensions, and that all parts have been installed. (Courtesy of Saab.)

Robot inspection and testing may range from simply placing a part in a fixture for resistance testing to sophisticated laser or optical gauging.

Robots used for inspection require the greatest accuracy of any application. It is estimated that 25% of inspection robots have a repeatability capability of ±.001 inches. By 1990, 12% of inspection robots are forecast to have a repeatability of ±.0001 inches.

The following sections present some applications of industrial robots for testing and inspection.

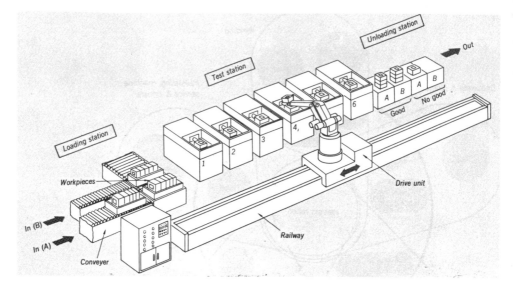

Fig. 6.4 Complete inspection line utilizing a traveling robot. (Courtesy of Hitachi.)

6.8 AUTOMOBILE BODY GAUGING

The use of noncontacting measurement instruments applied to robotic inspection has found great advantage in the automotive industry. An instrument frequently used is the Selcom Optocator by Selective Electronic, Inc. (Valdese, NC). This measurement is made by projecting a small beam of light onto the surface to be measured and then very accurately determining the distance to the resulting spot of light. When placed on the end of a precision industrial robot, the Optocator can be used to check critical dimensions of stamped, forged, cast, machined, and assembled objects.

In a typical application, the object to be measured would be placed at a known location relative to the inspection robot. The robot, having been previously programmed from a master part of theoretical calculations, moves to predetermined points in space around the object. At the points of interest, the Optocator takes distance measurements to the surface of the object. These measured values are then compared in the system against those taken from the master, or from the theoretical values. In this fashion, the dimensions of the object can be checked to see if it has been manufactured or assembled to within the desired tolerances. These measurements are also useful in determining trends in changes of the tooling or manufacturing process.

The Optocator system has resolution capabilities of 0.0003 inches. However, the overall measuring accuracy of the robotic inspection system will be a function of all components combined. Usually the limiting factor is the repeatability and accuracy of the

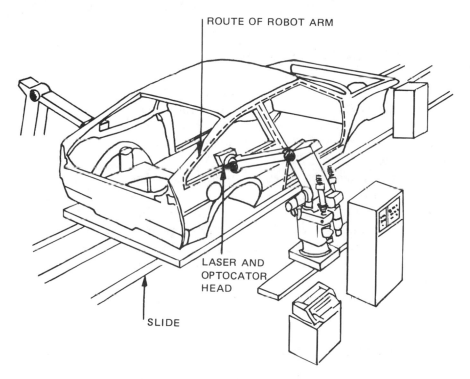

ROUTE OF ROBOT ARM

LASER AND OPTOCATOR HEAD

SLIDE

Fig. 6.5 Dimensional inspection of an automobile robot is accomplished by a robot-mounted laser gauge. (Courtesy of ASEA.)

robot. Typically, overall system accuracies of better than 0.004 inches can be achieved using commercially available electric robots, like the ASEA model IRb-6 robot pictured in Figure 6.5.

An automotive application is shown in the diagram of Figure 6.5. Two ASEA IRb-6 robots are used in this installation. The robots are mounted on either side of a shuttle line on the slides. This line can divert either all or a selected number of car bodies. This check is accomplished prior to the finishing of the body ("body-in-white"). The bodies are shuttled into position and shot-pinned in configuration of the doors, windscreen and backlight openings. The bodies can have 12 points per side gaged at a line speed of 90 bodies per hour (2160 points per hour). If a single car is to be checked completely (between 400 to 600 points) the cycle time would be approximately 14 minutes.

6.9 TCM BOARD TESTER

Figure 6.6 shows the IBM "Poughkeepsie Robotic Tester" which uses an IBM 7565 robot to test the electronic networks within TCM boards. The boards are used in IBM mainframe computers, such as the 3081 and 3082. Test probes are mounted on two robot arms which move independently.

The testing starts when the operator keys in the serial number of the board to be tested. The board is inserted vertically, and the position is established by the identification of three location points. Two test probes are sent to probe "C" screens to test if each segment of the board is good or bad.

The test of the TCM board by robot requires 2-1/2 to 5 hours. Previously, manual inspection required about 100 hours. IBM has 11 of the robotic testers in use at the Poughkeepsie plant.

6.10 VALVE TESTING

The task of manually loading and unloading leak-testing fixtures with gate and globe valves was found to be repetitive, boring, and

Fig. 6.6   IBM Robotic Tester used to test the electronic networks within TCM boards. (Courtesy of IBM Poughkeepsie.)

Fig. 6.7   Industrial robot prepares to load valve into one of a bank of test consoles. Use of two grippers avoids an extra trip to put down one valve before another can be picked up. Details of test fixture (inset) are shown with tested valve being unloaded. (Courtesy of Unimation.)

unchallenging by workers at the Cincinnati plant of Lunkenheimer, a division of Conval Corp. Now the job is being done by two industrial robots—each serving four test consoles—and the company reports lower costs and improved productivity (see Figure 6.7).

These benefits are due in part to incorporation of continuous production flow rather than the previous batch type operation which was characterized by skid-loads of valves waiting to be processed. The robot-fed testing operation is part of an integrated system which receives valves directly and continuously from the preceding operation, and, after testing, delivers acceptable valves to final assembly and packaging operations.

Valves to be tested are placed in the fixture on top of the test console with inlet, outlet, and stem bonnet aligned with three pneumatic clamping and testing cylinders. When the robot has withdrawn its hand from the test fixture, the testing cycle begins: pneumatic clamping and testing cylinders advance to seat firmly against the stem bonnet, inlet, and outlet surfaces of the valve shell. Air pressure checks seat and shell integrity. A manometer and photocell detect air-pressure differential that indicates valve quality. Valves that fail are automatically deposited into reject chutes designated as "defective stem" or "defective seat."

While the test sequence progresses at some of the consoles, the Unimate robot unloads a tested valve and loads an untested valve into the next console in sequence.

One of the robots at Lunkenheimer is responsible for testing valve sizes from 1/4 through 1 inch. The other one tests sizes from 3/4 through 2 inches. Changeover from a run of one valve size to another has been simplified so that no reprogramming is needed and set up takes only 5 to 10 minutes. The robot's fingers are changed to ones dimensioned to compensate for differences in body length and inlet/outlet diameters. The vertical and horizontal settings of the pick-up point on the delivery chute are adjusted to eliminate the shift in valve inlet/outlet centerline that occurs with changes in shell diameter. These alterations alone are sufficient for switching from one valve size to another,

6.11 X-RAY TESTING

The automotive industry has been using cast steering knuckles in ever increasing number primarily because of their reduced manufacturing cost. One drawback in using cast materials is the possibility of shrink holes being hidden within the casting which could seriously affect safety when used in the final application. Therefore, a very thorough test utilizing X-ray technology to find these weakening defects was developed for use at the Bergishe Stahl Industries, a Thysen Foundry in West Germany.

Each part is tested as it leaves the assembly line. A Reis robot picks the knuckle from one station, rotates it through a series of complex maneuvers, and deposits it outside the critical X-ray area. Each face of the casting must be thoroughly scanned by using X-rays to ascertain the necessary quality levels.

A six axis Reis Robot Model 625 was chosen for the task (see Figure 6.8) and installed inside the X-ray cabinet which is constructed of lead to keep the operator safe from harmful radiation. Each test

Fig. 6.8 Robotic X-ray testing. (Courtesy of Reis Corporation.)

position is crucial and to avoid the effects of over radiation or X-raying in the wrong position a stop is inserted into the X-ray beam in order to only and accurately display the exact positions needed to perform the test. The precise locations of the test positions may not vary any more than .019 inches from the reference positions.

Making accuracy control even more difficult is the fact that the parts are cast to relatively large tolerances. The part is tested in 11 different positions with one test being performed while the part is moving through a 90 degree rotation. The complete testing takes approximately 80 seconds. Included in that time period are increments of 4 seconds each per individual test in order to monitor the test results.

To insure that only faultless parts are used, reject parts are automatically marked with red paint to set them apart. This operation is performed merely by depressing a single control button on the operator's console.

All manipulations accomplished by the robot, and the test picture of the component, are displayed on monitors inside the operator's room. Programming and start-up are done remotely via a camera and monitor.

By utilizing a robot in this application, the capacity, the number of test positions, and the number of actual tests that could be performed in much less time were all significantly increased. At the same time, a marked decrease in the test cost per individual part was obtained.

To protect the operator from any possibility of radiation exposure safety gates at both the load and unload stations were installed. In order to grasp the specially cast knuckle inside the safety gate area a custom developed gripper is employed. Tooling in the checking area is made of Teflon to assure accurate results.

6.12 DETECTOR ALARM CHECK

The final electrical testing of smoke detectors by a United States Robot Maker 100 is shown in Figure 6.9. This inspection is the final station of a completely automated assembly cell which utilizes three

Fig. 6.9 Final electrical testing of a smoke detector. (Courtesy of United States Robots, Inc.)

Fig. 6.10 Inspection of electrical contacts by a Seiko RT-3000 robot.

Maker robots. The operation is controlled by a Square D CY-MAX 300 programmable controller.

To perform the final inspection of the detectors, two test probes apply 9 volts to the pallet mounted, assembled smoke detector. The robot activates a test button and listens for the detector alarm through a microphone. The robot either accepts or rejects the smoke detector by directing it to the proper chute. Smoke detectors which pass the test proceed on the pallet to packaging.

6.13 ELECTRICAL CONTACT TESTING

The Seiko Model RT-3000 robot is shown in Figure 6.10 in the inspection of electrical contacts. The robot first picks up the contact

Fig. 6.11 Testing of spring-loaded compliant contact by an Everett/Charles robot.

and performs a dimensional inspection by means of gripper sensors. If the contact is within predetermined tolerances, the robot proceeds to test the thickness of the gold plating by inserting the contact into an x-ray test chamber.

The testing of a more complex contact by an Everett/Charles robot is shown in Figure 6.11. The robot receives the spring-loaded compliant contact from a vibratory bowl feeder and passes it through test stations, including a resistance measurement circuit. The results of the test are displayed on a CRT and stored by computer.

7

Software for Quality Assurance

7.1 Operating System
7.2 How to Select Software for Automated Quality Assurance
7.3 Verifying Software Reliability
7.4 Directory of Automated Quality Assurance Software
Bibliography

The American National Standards Institute (ANSI) defines software as "A set of computer programs, procedures, and possibly associated documentation concerned with the operation of a data processing system, e.g., compilers, library routines, manuals, circuit diagrams."

All software may be divided into one of three categories: system software, applications software, and programming languages.

System software is comprised of programs or routines that belong to the entire system rather than to any single programmer and that (usually) perform a support function.

Application software refers to programs that let a computer perform specific tasks.

An assembler is a computer program that prepares a machine language program from a symbolic language program by substituting absolute operation codes for symbolic operation codes and absolute or relocatable addresses for symbolic addresses.

A compiler is a program that prepares a machine language from a computer program written in another programming language by making use of the overall logic structure of the program, or generating more than one machine instruction for each symbolic statement, or both, as well as performing the function of an assembler.

7.1   OPERATING SYSTEM

A computer's operating system is the key software "transportability." If two different brands of computers can run the same operating system, then many times they can both use a program written to run under its control. The operating system is software which controls the execution of computer programs and which may provide scheduling, debugging, input/output control, accounting, compilation, storage assignment, data management, and related services.

There are so many operating systems for microcomputers on the market today that it can be very confusing when trying to decide if a particular computer will run a particular program. Most major software packages have versions available to run on several operating systems.

In the past, operating systems, the software which controls execution of applications programs, tended to be specific to a machine or group of machines. Each computer vendor had its own. The "advantage" was the user had to come back to the vendor or his licensees to buy programs. Competition and the need for a unique marketing edge were also factors. But the rapid proliferation of microcomputers has changed the way users think, making software portability almost mandatory.

CP/M, developed by Gary Kildall, was the first popular microcomputer operating system. Because of the popularity of the IBM PC, the MS-DOS and PC-DOS operating systems currently share a position of importance for microcomputer software. CP/M is essentially for 8-bit machines, while PC-DOS is designed for 16-bit units. It appears that AT&T Bell Laboratories' UNIX operating system will be the base standard operating system for 32-bit micros and quite possibly for minicomputers as well.

As microcomputers become more powerful, the UNIX operating system developed by Bell Laboratories is emerging to bridge the gap between micros and minis. A growing number of vendors have brought out (or plan to) UNIX-based operating systems. However, since AT&T's licensing agreement with UNIX vendors expressly forbids use of the UNIX name, each company has its own. In fact because of some perceived weaknesses in the UNIX system, many companies have added their own specific enhancements. Hewlett-Packard, for instance, calls its version HP-UX. Microsoft's XENIX operating system runs on the new IBM AT microcomputer, as well as MS-DOS.

7.2 HOW TO SELECT SOFTWARE FOR AUTOMATED QUALITY ASSURANCE

Until recent years, programmers and engineers have traditionally written most software in-house. In the 1970s, a wide spectrum of quality software packages became available for a wide variety of business applications. As the computer industry matured, applications in manufacturing began to become developed as off-the-shelf packages. For any application, the computer user must assess carefully the availability and benefits of purchasing such software packages. This chapter provides guidelines to assist the user in selecting appropriate software based on the experience of software consultants who have viewed a wide range of applications.

The decision whether to develop software in-house or to purchase a standard program that is offered as an off-the-shelf package is a fundamental issue in developing any computer application. The first question is to determine if any software may exist for the application. If the computer user is locked into an operating system which is not rich in available software or the application is very specialized, the alternative may not exist to purchase a standard package. Where both in-house and commercial package options are available, the user must weigh the pros and cons of each approach. Ken White, a Senior Consultant with the management consulting firm of K.W. Tunnell Company, Inc., King of Prussia, PA, offers the following comparison of internal development of manufacturing software and the purchase of commercial packages (White, 1983).

The advantages of in-house software advantages are:

Designed specifically for the company
Easier to customize
Data processing acceptance
Cash flow

The disadvantages are:

Longer lead time
Time difficult to predict
Cost difficult to estimate
Risk of poor performance
Lack of user involvement
Turnover can negate internal experience

The advantages of purchasing commercial manufacturing software packages are:

Proven capability
Versatility, flexibility
Lower cost
Better documentation
Vendor supported maintenance

The disadvantages are:

Sometimes not feasible
Not easy to customize

Sometimes data processing resistance
Cash flow

Currently, most manufacturing firms tend to develop their own inventory and production systems rather than use commercial software packages (Guttman, 1984). Manufacturing systems must cope with complexities similar to those of wholesale and retail inventory systems, but even further magnified. Such systems must also separate inventory into finished goods, work in progress, and raw materials, and must provide a means of including labor in the cost of goods produced. Customer back orders and reorder points are affected by the manufacturing process itself; many components are assembled from parts that may themselves have to be assembled. This links the inventory system with production scheduling, which is likely to vary substantially among firms even within the same industry. With the recent advances in applications software from independent firms, there is clearly an increasing trend toward the use of software packages. International Data Corporation forecasts that independent software will grow over 10 percent in the next five years, while customized software will decrease by over 10 percent (Richter, 1984).

7.3 VERIFYING SOFTWARE RELIABILITY

The issue of reliability of engineering software is an important one. Untried software can cause safety problems as well as overruns in costs and scheduling. A program's reliability must be measured by the deviation of its test results from those of an independent source. A panel discussion on Quality Assurance for Engineering Software that was held at the 1983 ASME International Computers in Engineering Conference.

The methods used for computer programs are (Richter, 1984):

Independently verified computer programs
Hand calculations
Mathematical solutions
Empirical data

Technical literature
Comparison of:
    mathematical models (equations)
    numerical algorithms
    assumptions
    limitations

Evaluation of test results may be based on:

Accuracy of comparison method
Comparison of results
Deviations and their significance
Compliance with codes and industry standards

To avoid inaccurate results from engineering programs, government agencies, technical societies, and companies are active in the production of guidelines and standards for verification. In an article on the need for computer standards, Schuster (1981) reported the existence of 60 standards over 300 projects to develop more of them. Some of these efforts are limited to a series of definitions or brief directives.

7.4 DIRECTORY OF AUTOMATED QUALITY ASSURANCE SOFTWARE

Descriptions of commercially available software application to inspection and quality assurance are given in Appendix 2.

BIBLIOGRAPHY

Guttman, M.K. (1984). "How to Choose an Accounting Package," *PC World*, (Oct.), pp. 57-66.

Richter, H P. (1984). "Verifying the Reliability of Engineering Software," *Mechanical Engineering*, (Jan.), pp. 53-56.

Schuster, D.J. (1981). "Computer Standards: The Need and Challenge," *Mechanical Engineering*, (Feb.), pp. 40-45.

White, R.W. (1983). "How to Select Manufacturing Software Packages," Proceedings of the American Production and Inventory Control Society.

8

Inspecting the Product

8.1   Vision Inspection at General Motors
8.2   Vision Inspection at Kodak
8.3   Vision Inspection at Saab
8.4   Sheet Metal Inspection
8.5   Automatic Inspection of Engine Blocks
8.6   Pharmaceutical Inspection
8.7   Quality Assurance for Semiconductors
8.8   Automated Printed Circuit Board Inspection
8.9   Nonwoven Materials Inspection
8.10   Casting Flaw Detection
8.11   Glass Tubing Inspection
8.12   Other Applications
Bibliography

It has been estimated that 90 percent of all industrial inspection activities requiring vision will be done with computer vision systems within the next decade. Machine vision for inspection, however, is not just a technology for the future. Immediate, practical applications for automated inspection devices in U.S. industry are estimated in the hundreds of thousands.

This chapter presents applications of machine vision for the automated inspection of a wide range of products which have been demonstrated based on information provided by users and suppliers of machine vision systems.

8.1 VISION INSPECTION AT
GENERAL MOTORS

General Motors is reported to be using more than 500 machine vision systems, and has identified 44,000 potential in-house applications. Subsidiary GMF Robotics offers a commercial vision system, and GM has invested in five vision companies: Automatix, Diffracto, View Engineering, Robotic Visions Systems, Inc., and Applied Intelligence Systems, Inc. At its Chevrolet Division, a system is inspecting front-end stampings (2 seconds/piece) to see that each part has its full complement of 90 holes. Before the system was installed by Automatix, Inc., production problems could be identified only at the hourly inspections—which could allow over 1000 bad parts to be produced. In robot guidance, a Canadian GM plant is using the company's Consight vision system to sort castings. At Pontiac, MI, another GM-developed system called Keysight is checking for the presence of valve-spring retainer keys in a Pontiac Motor Division engine plant. The Cadillac plant at Livonia, MI, is using an Applied Intelligent System's unit to inspect cylinder heads, timing gears, and valve ring retainers. The GM Engine plant at Moraine, OH, is checking precombustion chambers; the Willow Run GM assembly division plant at Ypsilanti, MI, the mounting of tires.

8.2 VISION INSPECTION AT KODAK

Kodak has documented successful performance of more than 125 vision systems for more than a decade. For example, they installed a vision system in the packaging area of the Kodak disc product to reject any film cartridge with film slide improperly positioned. The system is affixed in front of the final packaging station, not so much to avoid packaging paper waste, but to prevent a slightly imperfect film cartridge from dropping into the consumer's camera. The liability factor of the latter is far higher. Machine vision also plays a significant role in front-end development of the disc film cartridge. Using optical systems, Kodak can perform 180 measurements in 15 minutes on the steel molds used to make disc cartridges. Previously, it would take weeks to manually measure only 15 dimensions, and several months could easily pass before part certification. Videk, a Kodak subsidiary, now commercially sells machine systems for industrial applications.

8.3 VISION INSPECTION AT SAAB

The interest in automated quality assurance has not been restricted to manufacturers within the United States. Several European automotive companies have installed machine vision systems and other computer-based inspection systems. Saab, for example, has developed an in-house machine vision system which is used in a wide range of applications. The system, designated the EVS 300, can automatically carry out advanced measurements, identification, or product inspection. It consists of one or two cameras which transmit the image data to a computer. The system can be combined with other control systems for continuous peripheral functions, such as conveyor belts, machines, etc.

Two applications are shown in Figures 8.1 and 8.2. The first figure shows the inspection for embedded components. The height, width, and installation depth are measured. Cracks in the plastic part and damage to the edges are detected. More than one component a second can be inspected.

Figure 8.2 shows turned surfaces on a flywheel being inspected by the EVS 300 to check that no casting pores occur. Defects down to one mm^2 are detected. The inspection station is included in a work center in which rough turning, fine turning, and drilling are carried out. A robot transfers the flywheel between stations.

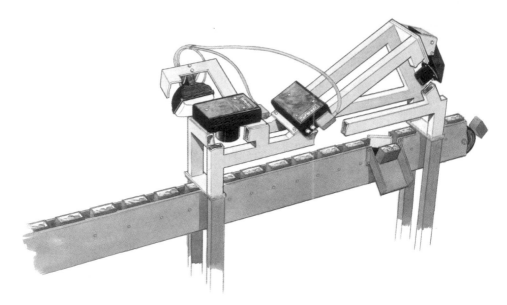

Fig. 8.1 Inspection of embedded components by the Saab EVS 300 machine vision system. The height, width, and installation depth are measured. More than one component per second can be inspected.

8.4 SHEET METAL INSPECTION

Metal stampings are subject to cracks, breaks, and tears in certain areas due to the stresses induced by the drawing operation. Many stampings are quite complex, making operator inspection time-consuming and complex.

A major automotive manufacturer installed the 15-camera system shown in Figure 8.3 on a progressive stamping line for the inspection of van innerside moldings and hood liners. Previous hand inspection procedures were time-consuming and labor intensive. One day it took three personnel an hour and a half to find the fault causing a missing hole on a panel. Now, the machine vision system by Itran Corporation (Manchester, NH) provides on-line inspection for 100 percent of the parts run. Inspection time is only eight seconds per panel. The installation was designed and installed by Martin Systems (Lansing, MI).

Fig. 8.2 Turned surfaces on a flywheel are inspected by the Saab EVS 300 to check that no casting pores occur. Both sides are checked in less than 20 seconds. Defects down to 1 mm² are detected. The inspection station is included in a work center in which rough turning, fine turning, and drilling are carried out. A robot transfers the flywheel between the stations.

The inspection of an automotive door panel by the Flexible Measurement System of Perceptron (Farmington Hills, MI) is shown in Figure 8.4. Some benefits of the system include: visual fixturing capability to compensate for workpiece mispositioning; flexibility of measurement and fixturing; a low-cost solution to replace expensive and inflexible fixturing; high levels of accuracy and repeatability

Fig. 8.3 Machine vision inspection of van innerside linings. (Courtesy of Martin Systems, Inc.)

while allowing for higher sampling rates and total statistical process control (SPC) reporting for process management decisions. As the process begins, the robot positions a vision sensor to determine a part's spatial placement. The system software compensates for part mispositioning and communicates data to correct the robot's pre-programmed path. The robot then positions the vision sensors above the first user-defined measurement point and activates the appropriate sensor. The measurement data is received and stored to ascertain compliance with defined specifications for future analysis.

Figure 8.5 shows the inspection of a truck cowling vent. While 100 percent inspection was required to insure proper hole placement

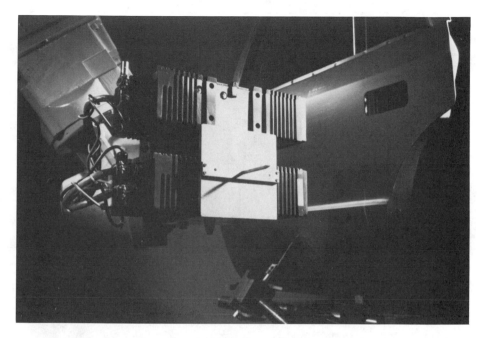

Fig. 8.4 Machine vision system measuring critical points on a door assembly. (Courtesy of Perceptron.)

and diameter for proper airflow, the large cowling was virtually impossible to hand inspect with any degree of accuracy and speed. The human mind simply cannot constantly focus on a part that has more than 1500 holes and accurately detect minor flaws. The cowling replaced standard louvers with 1580 holes pierced into galvanized steel. This design would keep debris from entering the engine compartment while maintaining the critical volume of cooling air flow. For cosmetic reasons, the loss of one hole made the part unrepairable, and a variance in the 4.2 mm hole size reduced the ventilation efficiency. The six-camera inspection station, utilizing an Itran machine vision system and installed by Martin Systems, performs the complete inspection within a 4.5-second cycle time.

These automated systems replace manual techniques where products are dimensionally qualified using off-line processes. These

Fig. 8.5   Machine vision inspection of over 1500 holes in a truck cowling. (Courtesy of Martin Systems, Inc.)

techniques range from simple feeler gauge measurement to sophisticated coordinate measurement machines (CMMs). Most of these methods have drawbacks, and do not necessarily provide cost effective solutions. Characteristically, the off-line techniques present challenges in process management, and the attaining of consistent measurements. The processes are slow, and measurement data is not timely for statistical process control (SPC). Qualification accuracy, with the exception of coordinate measurement machines, can vary between operators. Expensive fixturing is normally required for high measurement accuracy. From a management-control and cost-containment view, qualification methods that capitalize on

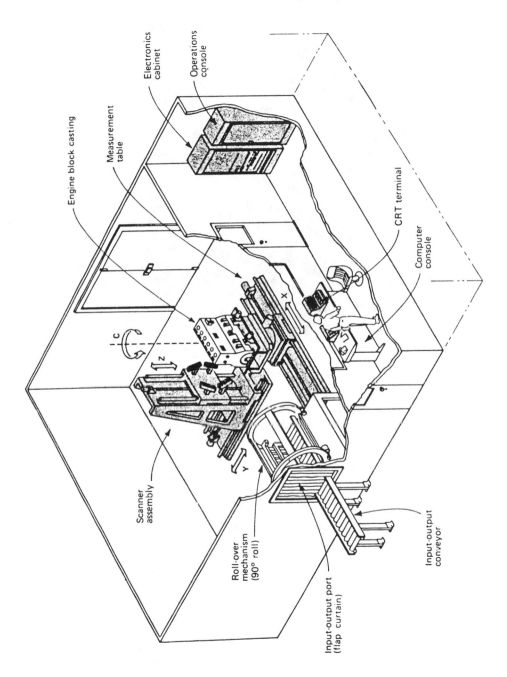

Electronics cabinet

Operations console

Engine block casting

Measurement table

CRT terminal

Computer console

C

Z

X

Y

Scanner assembly

Roll-over mechanism (90° roll)

Input-output port (flap curtain)

Input-output conveyor

Fig. 8.6   Three-dimensional vision inspection of engine blocks at Cummins Engine Company. (Courtesy of Robot Vision Systems, Inc.)

new technology must be used to obtain the desired product quality/ profitability goals.

8.5 AUTOMATIC INSPECTION OF ENGINE BLOCKS

A three-dimensional vision-inspection system was developed by Robotic Vision Systems, Inc., Melville, NY. An installation of the system at Cummins Engine Company for the inspection of engine castings is shown in Figure 8.6. Over 1250 points are checked in approximately 40 minutes. Manual inspection requires 40 hours. The system uses a laser beam light source and a triangulation algorithm

Fig. 8.7 Laser system for casting inspection using triangulation. (Courtesy of Diffracto, Ltd.)

to determine distance to the surface at any given point. A set of four fixed sensors scan portions of the surface until the entire surface is scanned. The casting is then rotated 90 degrees and the process continues. The final result is a three-dimensional computer map of the casting, which is then compared with a blueprint pattern stored in a computer to determine errors in dimensional tolerances.

Another engine casting inspection system is shown in Figure 8.7. The laser sensor utilizes triangulation to achieve a high degree of accuracy using triangulation. Typically, the casting would be placed on a micrometer actuated plate to allow measurement of

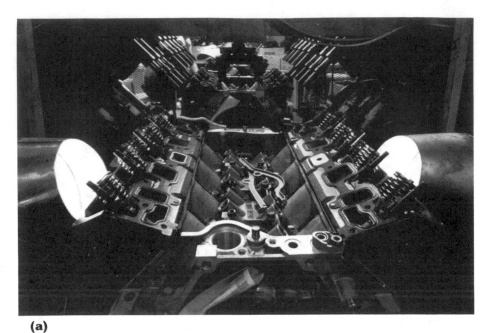

(a)

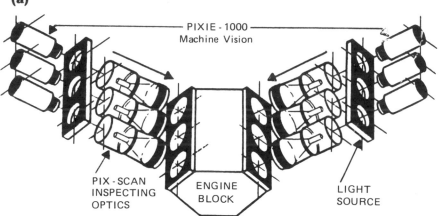

(b)

Fig. 8.8 (a) A machine vision system verifies the correct match of engine blocks and cylinder heads on an engine assembly line in photo. (b) Similar application of grading individual cylinder bores.

several points. While the laser probe could be manipulated by an industrial robot arm, the positioning repeatability of the robot would limit measurement accuracy. The Laser Probe is manufactured by Diffracto, Ltd. (Windsor, Ont.).

An automotive manufacturer installed the system shown in Figure 8.8 to improve the consistency and quality of its six-cylinder engines by grading each piston bore. The vision system shown is manufactured by Applied Intelligent Systems, Inc. (Ann Arbor, MI). A similar system is used to verify the correct matching of engine blocks and cylinder heads on an assembly line. Another automotive manufacturer uses the system to detect porosity in upper and lower ring grooves as well as in the piston bore area. Other applications which have been successfully demonstrated include seal inspection, groove inspection, and shaft inspection.

8.6 PHARMACEUTICAL INSPECTION

The real challenge for pharmaceutical equipment suppliers is in providing a 100 percent correct product to the consumer. The industry is striving to perfect means to reach this goal because the potential liability for companies regarding a product containing foreign material is just too high to compromise this search for quality.

Today, machine vision offers the greatest potential to the industry in achieving the goal of a 100 percent quality inspected product. Previous to machine vision, the only automatic detection possible was the absence of a product (short count bottles). With machine vision, conditions such as broken tablets, foreign objects, and the wrong product can also be detected.

An example automated quality assurance system is the Lakso Reformer 990 slat feeder (Leominster, MA). It has been equipped with LaksoVision, a technology supplied by Pattern Processing Technologies, Inc. (Minnetonka, MN).

Specifically what the Reformer 990 does is to spot missing tablets, damaged tablets (broken, split, etc.) with as little as 20 percent missing, foreign materials depending on size and orientation, and

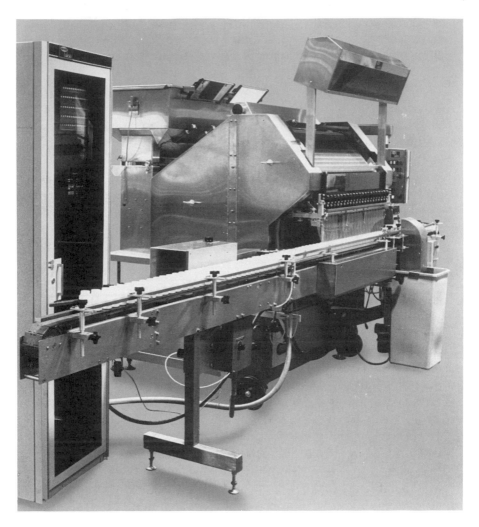

Fig. 8.9   Automatic slat filler with machine vision system. It will fill and inspect tablets or capsules at speeds up to 50,000 per minute.

foreign tablets depending on size. Once spotted, the defect is tracked to a given bottle and ejected from the conveyor at the eject station. The operational settings are key selectable from a library menu set-up and tested using the patented technique for vision data analysis that accounts for the system's high speed operation. The system is shown in Figure 8.9.

The LaksoVision system does its analysis in hardware circuitry in its Associative Pattern Processor (APP) from Pattern Processing Technologies, Inc. The way the APP works is to allow a camera scene to be divided into a series of small areas (windows) and then to be able to independently sense contrast levels in the windows. The technique is therefore called "windowing/contrast sensing." In simple terms, window/contrast sensing works as follows: (1) An image is obtained on camera of a "good" test item such as a number of tables in a slat filler. The test image must be visually correct in terms of what the vision system will look for to "pass" inspection. (2) The image characteristics that make it "good" are then boxed in with windows forming a good test template and labeled as such. (3) This "good" template is then compared to production line images and, in those which the contrast in the test windows does not match, the item is said to "fail" the test.

The application of the window/contrast sensing technique to tablet and capsule inspection in the Reformer 990 is very straightforward. There are five charge-coupled devices cameras mounted over the slats. Each camera inspects between 15 and 24 positions. The lighting is via a bank of infrared light-emitting diodes which are strobed to illuminate the slats at about a 40 degree angle. With each strobe, one or two tablets in each slat position are inspected by examining the reflected light in each tablet's inspection window. If the tablet is broken, not present, or the position is occupied by foreign material, the reflected light pattern will be different in the inspection window, thereby causing a change in contrast and indicating an inspection error condition. The simplicity of the technique is one of the primary reasons such high speeds can be achieved.

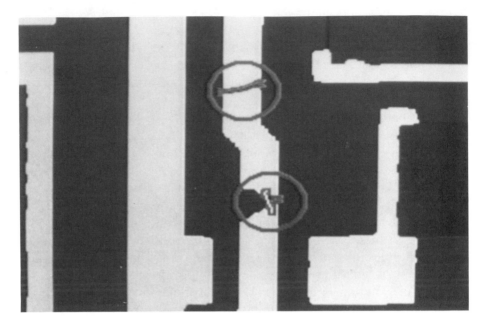

Fig. 8.10 Inspection for circuit flaws. (Courtesy of Synthetic Vision Systems, Inc.)

8.7 QUALITY ASSURANCE FOR SEMICONDUCTORS

Machine vision systems are now being successfully used in the semiconductor industry. Human inspection operators pose several problems as people unavoidably bring contaminants into the production area and are exposed to any dangers in the area. In addition, people are prone to error, cannot work at production line speeds, and are not cost effective.

Circuit flaws too small for the human eye to see are detected and reported by the TFI hybrid substrate inspection system from Synthetic Vision Systems (Ann Arbor, MI). Figure 8.10 shows the automated TFI system as it analyzes circuits for connectivity (top flaw shows a conductor break) and verifies that they meet design rules (bottom flaw indicates that circuit path width is less than minimum allowed).

Automated semiconductor wafer fabrication requires an accurate, fast, clean method for identifying and tracking individual wafer slices. Laser-etching techniques now allow manufacturers to mark identification characters on each wafer. To provide traceability, silicon suppliers laser-engrave codes into each raw wafer, and device manufacturers often mark the wafers with additional codes. Both the silicon supplier and the device manufacturer require 100 percent inspection to ensure that the marks are correct and legible. Cognex Corporation manufactures the Checkpoint 1100, shown in Figure 8.11. The Checkpoint 1100 is an intelligent vision system that can inspect the quality of wafer marks written in OCR-A, SEMI, or any other font. This system measures each character's shape, contrast, size and location. It also measures the character string's position and angle with respect to the wafer flat. It then determines the acceptability of the mark according to user-defined standards.

8.8 AUTOMATED PRINTED CIRCUIT BOARD INSPECTION

In electronic assembly manufacturing, circuit board loading errors can cause up to 40 percent of all product rejects. While those errors cost only a few cents to repair before the board is soldered, they can cost tens of dollars to repair later in the manufacturing process. Moreover, defects not found until later test stages limit the productivity of expensive, post-solder test equipment. Manufacturers clearly need fast, objective assembly test equipment for use prior to soldering.

The following summarizes some of the experimental achievements relating to printed circuit boards. The approach has been first to check for the presence and precise location of parts, then to pursue more detailed inspection for part damage and specific identi-

Fig. 8.11 Laser-etched characters in semiconductor wafers (above) are read by the DataMan 110 vision system from Cognex (below).

fication. The following domain-dependent constraints are used to aid in the inspection.

First, printed circuit boards are composed of relatively few classes of similar components that can be modeled with generic prototypes at several levels of specificity. The inspection algorithm for each class of components can be implemented with a small, modular procedure. That procedure can be invoked for each instance of a member of the class, thereby allowing a large inspection system to be built up from simple and nearly independent modules.

Second, the inherently two-dimensional structure of printed circuit boards implies that the shape of components is unaffected by position and orientation, even though the components may appear in different locations. (This, however, is not always true; components that project above the board on rather "floppy" leads may present different shapes to the camera.) The invariance of shape implies that components can be effectively modeled with simple, static geometric forms that appear unchanged in the image.

Third, printed circuit boards are organized with their components on a rectilinear grid. Because the orientation of the components is known, they need be located only in two dimensions, x and y. Furthermore, these dimensions are usually separable: one can first determine the x location and then the y location, or vice versa. This constraint implies that simple and very efficient one-dimensional signal processing techniques can be used to analyze intensity profiles from scan lines and scan columns.

An automated inspection system for the inspection of assembled printed circuit boards, offered by Intelledex (Corvallis, OR) is shown in Figure 8.12. The system simultaneously inspects a mix of device technologies for critical specifications such as lead length, clinch angle, wipe angle, and proximity of neighboring leads, as well as presence/absence and component polarity to provide a complete solution of assembled printed circuit boards before soldering.

Manufacturers using surface mounted devices (SMDs) face many of the same challenges as do manufacturers using conventional printed circuit board assembly methods: missing, wrong, misplaced and/or reversed components. The potential problem of

Fig. 8.12 The IntelleVue DR 2000 uses design rules to inspect populated printed circuit boards. It inspects lead length, clinch angle, wide angle, and distance between leads as well as component presence and orientation. (Courtesy of Intelledex.)

glue leakage onto the contact surface also exists. Due to the nature, speed, number of components and lack of maturity of the surface mounted device process itself, a large percentage of the boards often require some type of rework due to component placement errors.

The verification of surface mounted device placement by an Autovision system, from Automatix, Inc. (Billerica, MI), is shown in Figure 8.13. This system is available as a stand-alone unit, or with vision inspection built into automatic placement machines, as shown in the figure.

The inspection of PC boards could be considered to be extended to verify component correctness, such as by reading resistor bands. However, the lack of any standardization in electronics components precludes the inspection for the verification that an inserted

Fig. 8.13 Autovision system built into automatic placement machine verifies surface mounted devices placement. (Courtesy of Automatix.)

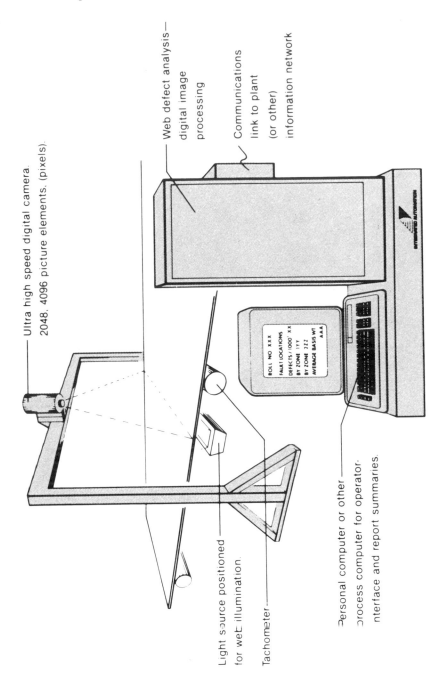

Ultra high speed digital camera.
2048. 4096 picture elements. (pixels).

Web defect analysis—
digital image
processing

Communications
link to plant
(or other)
information network

Light source positioned
for wet illumination.

Tachometer.

Personal computer or other
process computer for operator-
nterface and report summaries.

ROLL NO XXX
FAULT LOCATIONS
DEFECTS / 1000 XX
BY ZONE IYY
BY ZONE ZZZ
AVERAGE BASIS WT
AAA

INTEGRATED AUTOMATION

component is the correct one. Similar components manufactured by different suppliers vary to the extent that any attempt to inspect for component values would necessarily result in the rejection of good components because they may vary in appearance from the components which were used to teach the system. For this reason, all machine vision systems now inspect only for the presence of components, and do not check for correctness. It should be noted that components which have an asymmetric shape, such as some diodes, provide an exception to the limitation associated with component non-standardization, and can be checked for proper polarity insertion by a vision system. Fortunately, the insertion of incorrect components is a less frequent problem than missing components, and machine vision inspection offers a valuable service, even without verifying component values. Hopefully, future standardization by component suppliers will allow the development of this next step in PC inspection. Perhaps it will be the interest in utilizing computer vision component inspection which will provide the impetus for such standardization.

8.9 NONWOVEN MATERIALS INSPECTION

A vision system for inspection of nonwoven materials has been developed by Integrated Automation (Alameda, CA). The system, designated as Webspec 3000, is shown in Figure 8.14. It can inspect for missing components, spots, wrinkles, tears, holes, and variation in opacity or thickness (formulation index). Both defect location and size can be quantified.

The Webspec 3000 consists of a 4096 charge-coupled device (CCD) linear array camera, signal processing electronics and software and an attached IBM personal computer for report presentation. The system is mounted directly on the process line (base machine) or can be separately mounted on an attached slitter to inspect

Fig. 8.14 Webspec 3000 system for optical inspection of nonwoven materials. (Courtesy of Integrated Automation.)

continuous process of materials for various defects at speeds up to 3000 feet per minute.

The system allows for real-time inspection of these defects and can be used to activate markers or other alarm systems when a defect is detected. The marking system indicator is adjustable to allow for variations in the location of the marker as the system constraints dictate varying flagger or marker placements relative to the inspection location.

The system uses one or two CCD array cameras (4096 elements) to capture the image of the material as it passes under the camera. The material can be either back-lighted, if it is transparent and the defects relate to variations in transmission intensity, or it can be top-lighted to allow for surface defects which are not clearly visible with transmitted light. The light source is available in 120 inch or 70 inch standard widths but can be custom designed to allow for other cross deckle width variations.

The camera signal is (after a clocking signal from a process line tachometer) analyzed with a series of boards which provide real-time analysis of the signal. Defects appear as differences, high or low, in the signal intensity. The analysis circuitry presents the defect details to a signal board computer which analyzes the defects with an "object tracking" software program. The program collects information on sequential line scans (on the order of 1600 to 2500 line scans per second) and continues to track defects as single objects or multiple objects until certain conditions arise which signal the end of a defect.

The IBM PC is merely used to set inspection parameters in the inspection hardware and as a reporting device of defects as they occur on a split screen. The top half summarizes the defects by zone (1 to 16) and type (two levels dark or two levels light). The system has an option to separate defects to 16 levels of grey (grey level board and associated software). The attached printer prints data about defects as they occur to identify zone location and this information is also displayed on the bottom half of the PC screen.

The system can also be configured with infrared and ultraviolet sensitive camera and can be used to replace the backend electronics of the sensor technologies.

Fig. 8.15 High-speed flaw inspection of connecting rods using ultraviolet light. (Courtesy of Automatix, Inc.)

8.10 CASTING FLAW DETECTION

When a casting is illuminated by light from an oblique angle, it would be expected that a casting defect would cast a shadow which may be detected by a vision system. This technique, however, is found to be only partially successful in detecting larger flaws. Better results are obtained using a laser light, which diffracts at surface irregularities, or by using ultraviolet light.

An example of this class of application is the use of an artificial vision system with a classical magnetic dye penetrant and ultraviolet illumination technique. The forged automotive part shown in Figure 8.15 is magnetized, immersed in the dye, and then placed under ultraviolet illumination and inspected by the Automatix Autovision system to find dye that has seeped into cracks. The high-speed system can simultaneously inspect both sides of this forging at the rate of 60 per minute.

A research team at the School of Engineering, University of Bath, United Kingdom, is currently developing a robotic device capable of removing surface defects from die castings by abrasive machining. They have developed a suitable vision module incorporating a periscopic viewing arrangement attached to a solid-state array camera. Preliminary results suggest that defects can be detected using simple image analysis techniques, provided the surface geometry of the area being examined and the average intensity of the resultant image are taken into account.

8.11 GLASS TUBING INSPECTION

Object Recognition Systems, Inc., has recently demonstrated the suitability of the ScanSystem Model 200 to detect bubbles and other flaws in glass tubulation as it is being extruded. Specifically, the application calls for detection of these flaws as the tubulation is extruded at 54 inches per second. Detection is based on dark-field illumination. Only the light scattered by a flaw is observed by the camera.

8.12 OTHER APPLICATIONS

A list of additional machine vision applications for inspection was compiled by Gordon Vanderberg and Roger Nagel, while employed by the National Bureau of Standards, in conjunction with Kenneth White of Proctor and Gamble. Information on applications by private companies is often difficult to obtain. Often the very fact that a company is considering an application is proprietary. There are no details of the application available. However, the diversity of the list is, in itself, thought provoking:

1. grasshopper contamination of string beans
2. frozen food quantity count variations
3. dust detection of LSI substrates
4. bottle fatigue detection, to prevent explosions
5. thermometer inspection

6. label inspection on plastic bottles
7. golf ball label inspection
8. IC chip wire bonding
9. measurement of the effectiveness of a window defogger over time
10. automatic debarking, initial timber cutting, sorting of lumber
11. measurement of biscuit height so that packaging equipment does not jam
12. unthreaded nut detection
13. button inspection for the correct number of holes
14. sorting of porcelain seals by shape and configuration
15. automatic closed loop control of extruded gelatin sausage casing
16. hot roll steel web width control
17. particle inspection of pharmaceutical liquids
18. closure integrity inspection for pharmaceuticals

> is the rubber seal inside the cap provided by the supplier?
> is the cap tightened?
> is it cross threaded?
> is the seal over the cap properly in place?

19. blister pack inspection—is there one pill per blister?
20. china plate inspection for pinholes and bubbles on the plate
21. automated separation of whole almond nut meat from debris, shells, and damaged meat

BIBLIOGRAPHY

Gevarter, W.B. (1982). "An Overview of Computer Vision," National Bureau of Standards Report No. NBSIR 82-2582, (September).

Nitzan, David (1981). "Machine Intelligence Research Applied to Industrial Automation," presented at the 9th Conference on Production Research and Technology, Ann Arbor, MI, (November 3-5).

VanderBrug, Gordon J., and Nagel, Roger N., "Image Pattern Recognition in Industrial Inspection," NBS Report No. PB80-108871.

Villers, Philippe (1983). "The Role of Vision in Industrial Robot Systems and Inspection," presented at Electro 83, New York, (April).

Zuech, N. (1986). "An Overview of the Machine Vision Industry in Europe," SEAI Technical Publications.

Zuech, N., and Miller, R.K. (1987). *Machine Vision*, The Fairmont Press, Inc. and Prentice-Hall (copublished).

9

Inspecting the Package

9.1   Container and Label Inspection
9.2   Alphanumeric Character Inspection
9.3   Automated Closure Inspection
9.4   Pharmaceutical Product Counting

Containers generally have an importance beyond the obvious role of providing simple protection and easy handling of a product. Packages generally contain instruction on usage of the product (in the case of pharmaceuticals, these instructions can be of extreme importance) but in all cases, the instructions are important for customer satisfaction.

Each industry's product has its own unique packaging and information which may be contained on the package. In the food industry, for example, many products must be date coded. Packaged food products must also have some type of coding or information in order that suspected problems can be traced. In the electronics industry, components are coded. Missing or unreadable codes could result in the component being inserted into an improper circuit.

Because of the importance of this type of inspection, statistical sampling is often not acceptable. Furthermore, the high cost as well as potential operator error often prohibit operator visual inspection. Machine vision and other automated quality assurance systems have been shown to be well suited for this type of inspection.

Tamper-proof packaging is now being used in many industries— especially in pharmaceuticals. Inspection is often performed to insure that the tamper-proof device is in the proper state when the product leaves the manufacturing area. Again, automated quality assurance can provide a solution to this problem.

Following are some examples of the use of automated quality assurance for container and product packaging inspection.

9.1 CONTAINER AND LABEL INSPECTION

One of the most common applications for vision systems is for the inspection of containers (bottles and cans). The containers may be inspected for a variety of potential irregularities:

> missing, crooked or incorrect labels
> bottle fill height
> missing or damaged caps
> foreign can detection
> code dates
> flaw analysis for cracks, scratches, and voids

Two industrial installations of machine vision for container inspection are shown in Figures 9.1 and 9.2.

Where contrast is adequate, the system can be set up to detect horizontal and vertical translation errors as well as skew of ±1/8 inch on a label on the order of 4 inches on the side. Tears, foldovers, and labels can generally be detected by the system provided there is adequate contrast.

Fig. 9.1 Container inspection by machine vision. (Courtesy of Automatic Inspection Devices, Inc.)

Fig. 9.2 Container inspection by machine vision. (Courtesy of Eaton Corporation.)

Machine vision systems will typically work as follows in this application. At the beginning of a run of labeled products, an operator trains the system on labels whose appearance is acceptable. He may train on more than one container during this initial set-up, storing each input as a separate image in the system's memory. Training is accomplished quite easily following prompted instructions furnished by a handheld keypad entry terminal. At the conclusion of this brief training cycle the system is put in the run mode. In this mode, as containers pass the scanner label, images are captured and processed in an identical manner as in the training function, only this time "match/no-match" decisions are made by comparing the images generated by each of the sensors to their respective libraries. When a "no-match" is observed, the system automatically provides a reject signal which can be employed by the end user to activate a reject mechanism. If during the run, several containers are rejected that the line operator feels should not have been, the system merely has to be trained to accept those containers as well, thereby adding those images to the library created during the initial training mode. When dealing with a container that has two reasonable parallel label panels (front and back labels) the system can be provided with two sensors and lighting arrangements. It can handle up to four cameras and each sensor can be linked to its own independent library of images. Where label verification is as important as inspection, the system can be employed in a slightly different mode to guarantee that the front label and back label always correspond. Where this is a requirement, both camera outputs are fed into the processor simultaneously using a video splitter. In this way, the image of both labels is stored as a single image in the library. Consequently, any change in either label that corresponds to the above mentioned one percent change in the label's appearance will result in a "no-match" decision causing rejection of that container.

A number of companies have product offerings that specifically address these requirements of the packaging industry. Some, however, are limited in what they can perform because of registration compensation difficulties (especially with round containers), resolution limitations of the sensors, contrast difficulties that do not set labels off from containers adequately, detail that must be detected (some suggest to proofread labels), and acceptable variations in hue saturations and brightness in both the hues on the label and container.

9.2 ALPHANUMERIC CHARACTER INSPECTION

The manual reading of labels, lot codes, and other alphanumerics is a slow and boring task for human operators. Manual reading is also

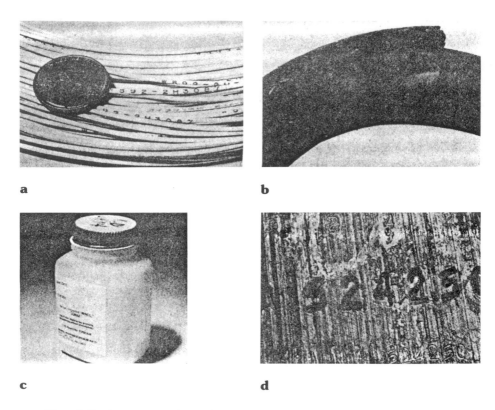

a   b

c   d

Fig. 9.3 Four examples of industrial alphanumeric reading: (a) aircraft wires, (b) rubber tire sidewalls, (c) pharmaceutical labels, (d) stencilled steel billets. (Courtesy of Cognex Corporation.)

error prone. For these reasons, machine vision systems are being utilized to automate inspection/verification functions previously requiring a human vision capability. Due to increased efficiency and 100 percent inspection, quality is improved and inventory control is more accurate.

The first industrial vision system for alphanumeric character recognition was DataMan by Cognex Corporation. There are also more of these systems in use today for alphanumeric information than any other model. Other companies offering character readers

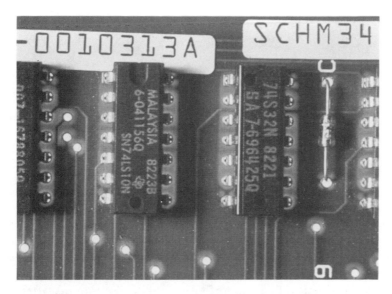

Fig. 9.4 Identification characters in integrated circuit packages can be inspected by machine vision systems. (Courtesy of Cognex Corporation.)

include: Key Image, EOIS, General Electric and Applied Intelligent Systems.

Four examples of alphanumerics which can be read in industrial environments are shown in Figure 9.3:

> hot-stamped characters on aircraft wires
> embossed, black-on-black serial numbers on sidewalls of rubber tires
> expiration dates and commodity codes on pharmaceutical labels.
> stencilled characters on steel billets

One of the largest cost areas in electronics manufacturing is the repair or scrapping of defective products. And in integrated circuit manufacturing, over half of those defects are errors in package printing—either incorrect or illegible identification codes. In the past, the only method of controlling integrated circuit print quality was by human inspection. But humans examine ICs much more

Fig. 9.5   Machine vision systems can inspect date and lot codes at production-line speeds. (Courtesy of Cognex Corporation.)

slowly than production lines manufacture them. So humans often sample products instead of inspecting every package, thus increasing the chances of defective products reaching the customer. This inspection is shown in Figure 9.4.

To ensure consumer safety and adhere to government regulations, most food and drug packages display expiration dates. Lot codes also appear frequently on packages in the food, drug, cosmetic and textile industries. Such codes, added to a package date in production, must be checked for correctness and legibility (see Figure 9.5). Until recently, manufacturers relied on humans to inspect date and lot codes. But human inspection is not cost effective. Also, such inspection is so slow that in many cases, checking every unit is impractical and packages on the line are merely sampled. Consequently, there is an increased chance that defective products will reach the consumer. Even when every unit is inspected, humans are still prone to err due to boredom and fatigue.

9.3 AUTOMATED CLOSURE INSPECTION

ZapatA Industries, Inc. (Frackville, PA) manufactures over 5 billion crowns per year—more than any other company in the country.

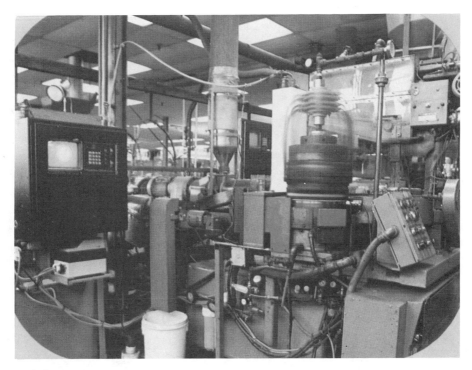

(a)

Fig. 9.6 Machine vision inspection on a closure production line. (Courtesy of ZapatA Industries.)

The firm uses the video inspection system shown in Figure 9.6 to inspect the sealing ring, corrugation form, center panel, and diameter of bottle caps. The system operates at high speed, up to 3,000 parts per minute, and allows 100 percent inspection of the company's product.

Results of the installation of this automated inspection system were a 66 percent reduction in the spoilage rate and a 500 percent increase in inspected product throughput as compared to the old manual inspection procedure. Quality of crowns improved more than 100 percent over the industry quality standard after the systems were installed.

The video inspection system was custom designed for ZapatA by Prothon Division of Video Tek, Inc. (Parsippany, NJ).

(b)

Fig. 9.6 (*Continued*)

9.4 PHARMACEUTICAL PRODUCT COUNTING

An automated inspection system for pharmaceutical in-line processes which require a high degree of accuracy in couting of a product has been developed by Automation Intelligence, Inc. (Orlando, Fl.). This system significantly increases counting accuracy and decreases total inspection time and labor over manual methods. In applica-

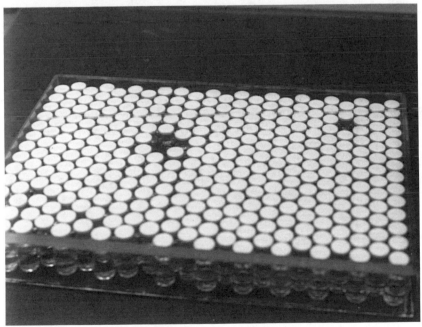

Fig. 9.7   Reconciliation system for pharmaceutical product counting. (Courtesy of Automation Intelligence.)

tions where the product is handled in a batch flow process, the reconciliation system can increase counting accuracies by a factor of 20 or more while, at the same time, decreasing counting times by a similar amount. The system has the capability to detect and count touching parts with an accuracy approaching 50 parts per million (0.005%). The gray-scale processing also allows the system to detect and count different product colors, diameters, and random positioning without manual intervention.

In operation, this integrated hardware/software system counts pharmaceutical products in batches as they move down a conveyor belt. A grouping of the product is collected and presented to the counting system following its placement in trays or other containment equipment. A photograph of the system in operation is shown in Figure 9.7.

10

Developing the In-House Program

10.1   The Automated Quality Assurance Plan
10.2   Implementing Machine Vision
10.3   Implementing Robotic Inspection
10.4   The Computer Communications Problem
10.5   Manufacturing Automated Protocol
10.6   The Technology of Manufacturing Automated Protocol
10.7   Other Computer Protocols and Standards
10.8   Computer Simulation
10.9   Simulating a Robotic Inspection Work Cell
10.10   How to Simulate an Automated Quality Assurance Project
10.11   Assessing the Benefits

A successful corporate automated quality assurance program must get off to the right start. A well-developed AQA plan is essential to achieving this. The first section of this chapter presents guidelines for developing the AQA plan, and implementation guidelines for machine vision and robotic inspection are discussed.

Computer protocols and networking standards must be observed to insure that the automated quality assurance computer system can exchange data with other manufacturing computers. Sections 10.4 thru 10.7 discuss the use of computer protocols to solve the communications problems between systems of different manufacturers.

Complex automated quality assurance systems should be designed and tested by means of computer simulation. This area is discussed in Sections 10.8 thru 10.10. Last, generalized guidelines for assessing the costs and benefits of AQA are presented.

10.1 THE AUTOMATED QUALITY ASSURANCE PLAN

Jack Conaway, from Digital Equipment Corp., identifies the following steps in implementation:

1. Create top management commitment
2. Find a project leader
3. Establish a corporate task force
4. Develop corporate standards
5. Perform modeling, analysis, and system design
6. Break the system up into smaller projects
7. Implement bottom up

Dr. Charles M. Savage, D. Appleton Company, Inc. (Huntington Beach, CA), provides the following check list for developing a "Strategy and Implementation Program":

Clear statement of business strategy and goals.
List of Critical Success Factors (CSF) to support the business strategy. What key things must go right to achieve this strategy?

Clear statement of strategy which supports business strategy and CSF.

CEO/GM charter for AQA Strategy and Implementation Program and top support from all functions.

Identification of the company's starting point.

Strong commitment to overall integration.

Survey existing systems to determine levels of compatibility. How are many software and hardware systems capable of communicating together?

Support from the engineering and manufacturing departments.

Agreement on standards and protocols.

Development of phased process to support the transition to AQA:

goals/plans
conceptual design
detailed design
implementation
benefits tracking and fine tuning

Ability to learn from and receive support from the hardware and software vendors.

Ability to learn from experiences of other companies.

Willingness to streamline or simplify existing operations. (Reduce the complexity index).

Use of outside resources (universities, professional associations, and consultants).

Identify potential financial benefits.

Develop new ways of working with and exchanging information with vendors and customers/clients.

Make use of public domain information from the various federal projects, such as NASA's IPAD, USAF's ICAM, and the Advanced Manufacturing Program of the National Bureau of Standards.

10.2 IMPLEMENTING MACHINE VISION

As with any technical system application, the cost, schedule, feasibility and competitiveness of a machine vision system solution are intimately tied to the requirements which are defined. Due to the complexity and relative newness of widespread vision system use, it is unfortunately easy to define specifications which are subject to misinterpretation or misunderstanding.

The Environmental Research Institute of Michigan has prepared the following list as a guideline to issues and factors which should be considered in developing the requirements for a machine vision application. Each item important in a particular application should be expanded as necessary and quantitatively defined based upon the real manufacturing and economic requirements faced by a user.

1. Cycle time or throughput rate per unit part or per volume of material
2. Repeatability (consistency of measurement in an unchanging situation)
3. Accuracy (consistency of measurement in an unchanging situation)
4. Resolution (smallest change measurable by system)
5. False reject rate versus false acceptance rate of vision system decisions
6. Variation in parts or materials to be processed and their placement
7. Variation in illumination and surface finish conditions to be tolerated
8. Variation of colors, orientations and textures on parts or surfaces
9. Variation of features, defects, and shapes to be recognized or measured
10. Degree to which parts and features may overlap or occlude each other
11. Constraints which exist on where sensors can be placed relative to the material or object (especially if application is on-line)
12. Minimum size of features or flaws to be measured or recognized
13. Size of largest part or material cross section to be analyzed

14. Future process or part changes which must be accommodated
15. Compatibility requirements with other automated equipment, communications, etc.
16. Environmental conditions to be tolerated—temperature, voltage, moisture, vibration, chemicals, etc.
17. Likelihood of requirement for future increase in capacity, functions, or interfaces
18. Skill and knowledge level of expected system operator
19. Form of data presentation and output information: graphic display, computer network output, statistics, process control, etc.
20. Cost required to achieve acceptable payback period
21. Number of systems anticipated

10.3 IMPLEMENTING ROBOTIC INSPECTION

There are several aspects of an in-house robotics program:

The feasibility study
Organization and planning
The first robot installation
Safety
Public relations

A survey of manufacturing operations is the first step in identification of a plant's potential robot applications. It is desirable to use a balanced four-person team consisting of a member of manufacturing engineering management, the foreman for the area being surveyed, an area manufacturing engineer, and a robotic consultant. In the plant survey, the team should assess both practicality and economics. The following is a list of important aspects to consider:

Simple, repetitive operations
Cycle times
Part location
Part weight

Inspection requirements
Number of shifts per day
Frequency of set-up

A manufacturing environment where robots would fit has three elements:

1. The products are hardgoods, which can be handled (picked up, put down, stacked).
2. The products are turned out in reasonable quantities, but not such high volume that hard automation would be the more practical way.
3. The plant operates with multishifts, so that the cost of the robot can be amortized quickly.

A successful first installation is probably the most important step in establishing an effective long-term robotics program. Cincinnati Milacron, a major manufacturer of industrial robots, has provided the following guidelines for the first robot installation:

1. Get management support and commitment. This includes assigning responsibility to a group, existing or newly created, for the purpose of investigating potential applications and implementing robots in the applications that show promise. It is important that at least some of the people in this group be familiar with the manufacturing areas and processes being evaluated for robot application. Also, they should be away from day-to-day operations so they are not putting out fires.

2. Make the first robot installation a relatively simple one: Walk before you run. If this first installation is successful, it will be easier to justify additional installations. Potential ROI should not be the primary concern. If the application with the greatest potential ROI is the most difficult in the plant, that potential means nothing if it is not successful. Valuable expertise is gained with initial installations. Expertise gained from early simple installations can be applied later to the more difficult ones.

3. Communicate with the shop people. Don't spring any surprises. Advance notice that a robot is on the way will help make the transition a painless one, for everybody.

4. Allow time for proper training and project debugging.
5. Take advantage of the support offered by the vendor.
6. Remember human engineering.

Vernon E. Estes, General Electric Company, developed a list of rules of thumb when applying robots. The list reflects GE's practical experience in robotics, and was compiled to benefit people thinking about using this form of automation to help their business. Estes' rules of thumb when applying robots are:

1. Don't try to implement robots or any other new technology without properly trained personnel. Like flying with an untrained pilot, the ultimate project direction is down.

2. Make sure you have your priorities straight. Which project should be tackled first is important, but even more important is education before tackling implementation.

3. Keep it simple! Don't buy a real-time seam tracking arc welding system if your parts are precise and through simple fixturing acceptable welds could be produced without it.

4. Robot uptime can be as much as 98 percent, but robotic system uptime will be less. Every component of any system has an associated downtime. No one can realistically promise 100 percent uptime. The efficiency of an automated system and the speed of its components will increase workstation output.

5. Select a full service robot vendor. This could be the difference between project success and failure. Will the vendor train your personnel (both pre-purchase and post-purchase), provide robots and related hardware integrated into a system, and then service what they sell? Will the vendor have the staying power while the industry is going through its maturing stages?

6. Treat any robotic system justification and implementation just like any other capital equipment investment, but make sure you call them "robot systems."

7. The robot is a fraction of the cost of the typical robot system. Low technology robots are typically 1/4 to 1/6 of the system. Medium technology robots are on the order of 1/3 to 1/4, and high technology robots are 1/2 to 1/3 of a system cost.

8. Robots must be applied as an integral part of a system. Failure is inevitable if the robots are not tied into the environment through the use of sensors.

9. The best way to find potential robot applications is to discuss chronic medical problems with your company nurse. Look for areas producing repetitive injuries of any type, or areas producing chronic muscle aches and pains.

10. Robot system safety must be the number one priority. The robot work area must be controlled with physical barriers. The human movement in relation to the robot work area must also be controlled by restrictive barriers, such as fences.

10.4 THE COMPUTER COMMUNICATIONS PROBLEM

The single greatest problem in the development of full-scale computerized manufacturing or process systems is that computer systems by different manufacturers (or often by the same manufacturer) seldom can communicate with each other. Data measured for one purpose cannot be shared for a higher purpose. This is necessary for any system to accomplish plant-wide automation.

The local area network (LAN) permits the user to distribute processing functions across several processors to increase performance, expandability, and reliability. Standardization of LAN communication protocols is a necessary key step in the development of computer-inegrated manufacturing.

Major computer users and vendors are working together with standards organizations to help define standards for communication at higher plant levels. The leading effort is the IEEE-802 committee. The IEEE-802 standard will provide a common communication network architecture that will link a wide range of manufacturing control units, data manipulation devices, and management data bases. The National Bureau of Standards, ANSI, ECMA, and the ISO are also involved in local area network standardization efforts.

General Motors, a user of over 20,000 programmable devices, has developed the manufacturing automation protocol (MAP) specifications as a standardized local area network approach. This effort is supported by the NBS, numerous computer system vendors, and 70 major users. The first manufacturing automation protocol demonstration was at the 1984 National Computer Conference, with General Motors receiving multivendor cooperation from Concord Data Systems, Hewlett-Packard, Digital Equipment, Motorola, IBM, Gould Modicon, and Allen-Bradley, for the MAP specifications demonstration program. The other companies making programmable controllers, General Electric, Reliance Electric, and Square D, have announced support for the MAP specifications.

General Motor's actions have been welcomed by virtually all industry leaders in manufacturing and electronics, since they, too, have nothing to gain and much to lose from an unstandardized market. The manufacturing automated protocol program has been endorsed by IBM, Intel, DEC, Ford, Allen-Bradley, Gould, Boeing, Westinghouse, GE, Kodak, Dupont, Volkswagen, Renault, and many others. All of them belong to a manufacturing automated protocol users group which is working on details of the program. A special panel of the IEEE—the 802.4 subcommittee—is developing MAP standards, as is the National Bureau of Standards. There is also support for MAP from Canada and Europe. Clearly, MAP is on the verge of becoming a global standard for industrial automation, and has given a major impetus to the field, bringing orderly growth to a market in a way that only standards can.

10.5 MANUFACTURING AUTOMATED PROTOCOL

Manufacturing Automation Protocol (MAP) is a communications standard being developed by industry to allow different types of computerized manufacturing equipment to communicate with each other. By using a common communications standard, the different "islands of automation" that now exist in many factories will be linked electronically into computer-integrated manufacturing systems (the automated, integrated factory).

The communications problem exists because different vendors make hardware processors and factory floor devices that may or may not be able to communicate with each other. In many cases, communications compatibility can be achieved only through costly custom interfaces.

A chronological history of manufacturing automated protocol is:

1980	General Motors forms an internal task force to begin work on the MAP program.
1982	GM makes first formal announcement of MAP.
March, 1984	More than 30 companies attend first MAP users group meeting, held in Detroit.
July, 1984	Portions of MAP are demonstrated at a National Computer Conference exhibit, where devices from IBM, DEC, HP, Gould, Allen Bradley, and other companies are linked together.
Fall, 1985	GM uses pilot MAP program in a Detroit-Hamtramck plant; MAP links programmable controllers in charge of spray-painting robots.

The turning point in recognition of the MAP standard was the November 1985 Autofact conference, attended by over 20,000. A quarter-acre exhibit sponsored by General Motors and Boeing demonstrated a MAP local area network linking computers and plant automation of multiple vendors for effective computer-integrated manufacturing. The MAP network was connected to the office local area network TOP to demonstrate connection of computers for effective office and administrative functions. A MAP Ethernet linkage was also demonstrated. Participating vendors demonstrated their ability to communicate across any and all of the following interconnected networks: IEEE 802.3/10 Mbs LAN, IEEE 802.4/5 Mbs LAN, IEEE 802.4/10 Mbs LAN, and through an X.25 Public Packet switched network to a remote IEEE 802.4/5 Mbs LAN.

10.6 THE TECHNOLOGY OF MANUFACTURING AUTOMATED PROTOCOL

As a specification development program, manufacturing automated protocol (MAP) is based on the International Organization for

Standardization's (ISO) seven-layer model for Open Systems Inter-connection (OSI). Specifically, MAP is described as a seven-layer, broad-band, token-bus based communications standard.

Each layer performs a set of functions necessary to provide a defined set of services to the layer above it and also interfaces with the layer below. These layers range from the physical layer at the bottom—involving the electrical, mechanical and functional control of data circuits—to the application layer at the top, covering the application process and management functions.

The MAP specification eventually will include international standard protocols for all seven levels.

Initially developed by General Motors as an internal specification for its manufacturing divisions to provide common communications, MAP technology has spread throughout manufacturing with several hundred companies now participating.

The seven layers of the ISO model are presented in Figure 10.1. They are:

Layer One, the Physical Layer: This layer deals with most of the hardware questions involved in networking, such as the actual medium that will be used and the rates at which data will be sent.

Layer Two, the Data Link Layer: This layer concerns the composition of the actual "frames" of data that are sent out on the network, as well as the protocols for getting those transmissions onto the network.

Level Three, the Network Layer: This layer sets up the procedure for communication between different networks.

Level Four, the Transport Layer: This layer makes sure a message reaches its destination. It is responsible for assembling a message into frames, then re-assembling the frames upon arrival and sending back a receipt acknowledgment.

Level Five, the Session Layer: At this layer, connections are established between stations, and logical names are mapped onto physical addresses.

Level Six, the Presentation Layer: This layer converts the message into a form that can be used by the receiving device.

Level Seven, the Application Layer: This layer brings network services, such as electronic mail, to the end user.

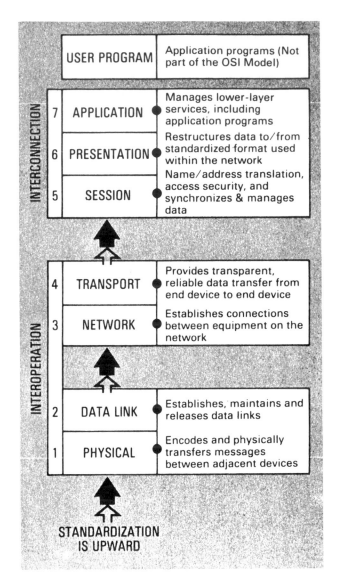

Fig. 10.1 ISO seven-layer model.

10.7 OTHER COMPUTER PROTOCOLS AND STANDARDS

Manufacturing automated protocol (MAP) is not the only proposed protocol for computer communications, nor does it address all issues. There exist several other networking standards, including some very recognized proprietary networks. The technical and office protocol (TOP) standard has been proposed for office communications, and IGES has been proposed for computer graphics. In addition, a variety of bus standards exist.

Boeing Company is trying to bring MAP-like standardization to technical office communications with its technical and office protocol (TOP), which was demonstrated at Autofact '85 in Detroit. TOP also follows the open systems interconnection (OSI) model.

Technical and office protocol is a specification for non-proprietary, multi-vendor data communications for use in technical and office environments. It is an application oriented subset of the OSI protocols and can be used in a wide variety of industries. The initial thrust addresses functions needed to integrate manufacturing and other office, technical, and plant floor environments.

Technical and office protocol is an evolving specification which currently reflects work accomplished in the National Bureau of Standards Workshop for Implementors of OSI Protocols and by the MAP efforts. Additional functions will be added to TOP as new work progresses.

The functions that TOP will address include:

Document, spreadsheet and graphics exchange (both revisable and final form)
Print, plot, file, and directory servers
File transfer (current capability)
Distributed database interfaces
Electronic mail, and store and forward messaging

Technical and office protocol shares a core of common protocols with MAP version 2.1. Because of this, a TOP and MAP network may be linked together and treated as a single logical network. Presently, the MAP and TOP protocols are exactly the same for layers 2 through 5. At layer 6, neither protocol has been specified. Layer 1 specifies the IEEE 802.3 CSMA/CD protocol. There will also be divergence at layer 7.

Both MAP and TOP have not been developed in a vacuum. One set of independent standards is completely defined and being used. Called the transmission control protocol/internet protocol (TCP/IP), this has been a Department of Defense standard since about 1978, and was officially adopted for all defense department networks in 1983.

Transmission control protocol/internet protocol is the result of many years of research by the Defense Advanced Research Projects Agency (DARPA). The ARPANET started with a four-node communications equipment report that many of their customers are following the TCP/IP protocol because it has been completely defined. In fact, many vendors who are trying to provide MAP and TOP products will still support TCP/IP.

Recent Department of Defense studies have suggested that TCP/IP networks be converted to International Organization for Standardization (ISO) protocols through gateways. This move toward ISO may be related both to the push in U.S. industry to ISO standards, and to the fact that European NATO computer installations will also use ISO standard networks.

Other protocols, such as the Xerox Network Services (XNS), are proprietary to one company but have gained some acceptance in the marketplace. Xerox Network Service is noteworthy, according to experts, because it was one of the first network architectures to try and implement the open systems interconnection (OSI) model. Some experts believe that as MAP standards develop, they will be incorporated into proprietary network standards such as XNS and Digital Equipment Corporation's Digital Network Architecture (DNA). In fact, DEC, has made a commitment to incorporate the MAP protocol, as it matures, into all DEC networking products.

There are several proprietary networks available that are used in industrial applications. High level networks include:

Ethernet (many sources)
TIway II (Texas Instruments)
Token/Net (Concord Data Systems)

VISTANET (Allen-Bradley)
Modway (Gould Electronics)
DECNET (Digital Equipment Corp.)

Some low level networks are:

Cop Net (ISSC, Honeywell)
GEnet (General Electric)
LAN 9000 (Hewlett Packard)
R/Net (Reliance)
Sy/Net (Square D)
WDPF (Westinghouse)
Sinec L1 (Siemens-Allis)

10.8 COMPUTER SIMULATION

A simulation is a model of a system. Such a model represents the importance of the performance of a system in a way which permits it to be easily manipulated. The main value of simulations is that they can be conveniently manipulated until the desired results are achieved.

It is possible to perform simulations by hand. Most of the simulations that are interesting, however, are of such size that manipulating the markers and keeping "score" becomes cumbersome. Thus, computers are used for virtually all manufacturing simulations.

The concept of using special computer languages for simulation dates back more than 25 years and the application of simulation for manufacturing dates to the early 1970s. Four factors have resulted in simulation techniques emergence as one of the hottest areas of industrial technology today: (1) availability of powerful microcomputers, (2) advances in computer graphics for simulation animation, (3) industry emphasis on flexible manufacturing systems, and (4) development of techniques to use simulation results for robot off-line programming.

Robotic simulation has developed from the robot time-and-motion techniques of the late 1970s to the current animation systems with automatic off-line programming capabilities for multiple robot work cells. In the design of robot work cells, up to 80 percent of the time required for implementation is spent on cell design and programming—this time can be dramatically reduced by the use of simulation.

Because of the complexity of flexible manufacturing systems (FMS)—which generally have some type of in-line inspection—and some material handling systems, computer simulation is almost essential. Multimillion dollar mistakes can be avoided. Simulation is frequently used for automated guided vehicle systems and industrial process design.

Because of the complexity and interaction of the various components, it is difficult to imagine how to predict how any manufacturing system will perform without simulation. The need for simulation for flexible systems is even greater because the system must handle a flexible demand in terms of product mix as well as volume.

Emulation is a simulation concept that adds a level of detail to a model. The computer uses the actual inputs and outputs from control devices (programmable controllers, minicomputers, etc.) to operate the simulated material handling system. Essentially, this allows the user to hoodwink the host computer into thinking that it is using the actual system. This approach provides a valuable tool in debugging control systems software. It provides numerical information that is often unavailable from a simulation study. Emulation is used to evaluate specific hardware after the handling and control systems have been designed. The emulation concept was developed by HEI Corporation (Carol Stream, IL) who received a patent for the first emulation system.

The tendency in nonsimulated design is that some elements are generally underdesigned and others are overdesigned. Based on user experience with simulation, it is generally found that a 20 percent improvement in the initial design of a system can result from simulation. The real economic benefit of simulation is that a workable manufacturing system is achieved at start-up instead of two to three years later.

The greatest paybacks for simulation are for projects with the greatest complexities. In flexible manufacturing systems, for example, the complex interaction of machine tools, robots, material handling systems and control algorithms make million-dollar mistakes easily. In an interview with Computer-Aided Engineering, a manager with the flexible manufacturing system (FMS) installer Giddings and Lewis commented: "Without a simulation, I wouldn't quote an FMS on a bet." This opinion is shared by virtually all of those in the FMS business—it is estimated that 100 percent of the systems being installed in the United States at this time have been simulated.

10.9 SIMULATING A ROBOTIC INSPECTION WORK CELL

Real-world industrial robots have a motion which is uniquely determined by their geometry or joint configuration. For accurate path projection planning, a simple velocity computed as a point-to-point distance divided by time does not provide an acceptable model of a robot. To allow users to design new robots, modify existing robots, and evaluate existing robots in their working environment, arms must be kinematically modeled as in Figure 10.2.

A computer model called GRASP (General Robot Arm Simulation Program) was developed by Stephen Derby at Rensselaer Polytechnic Institute to simulate motion elements of robot arms. In the model, robot displacement curves are constructed, assuming that all links are rigid. GRASP can be used to describe simulated work cells which are drawn on a vector display screen. As the robot moves about the workplace, the graphics display is updated. Figure 10.3 shows the GRASP model of a six degree-of-freedom robot transferring a workpiece from a conveyor to an inspection device. Figure 10.4 shows the resulting motion of the first actuator joint as the robot moves from the worktable to the conveyor.

The simulated robot motion can be specified in two ways:

1. The actual joint motions can be given. Working from the base joint out to the end effector with the joint motions allows you to find the unique position and orientation of the end effector.

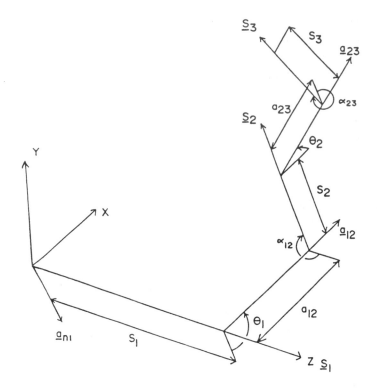

Fig. 10.2   Example vector definitions for kinematic modeling of robot arms.

2. The position and orientation of the end effector can be specified and an inverse kinematic solution can be used to find the joint values corresponding to the target position and orientation. The inverse kinematic solution may be a closed-form solution unique to that robot design or a numerical iteration algorithm that could handle a general 6-axis robot.

The task of designing a robot work cell can be reduced from several days to a few hours using computer-aided design (CAD) techniques based on models such as the one discussed above. In addition, the verification that the cell design is functional can be

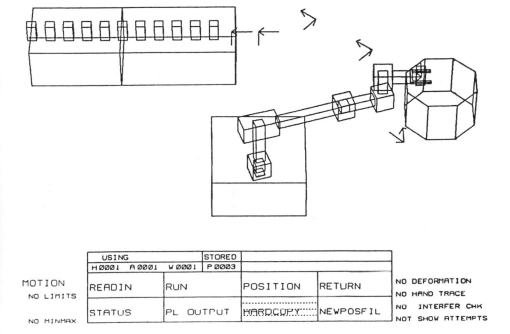

USING			STORED			
H 0001	A 0001	W 0001	P 0003			
READIN	RUN		POSITION	RETURN	
STATUS	PL OUTPUT		HARDCOPY	NEWPOSFIL	

MOTION
NO LIMITS

NO MINMAX

NO DEFORMATION
NO HAND TRACE
NO INTERFER CHK
NOT SHOW ATTEMPTS

Fig. 10.3 GRASP model of a general purpose six degree-of-freedom robot showing the robot's relationship within a workcell.

achieved on the CRT rather than risking that a design error will only be found during start-up of the actual equipment.

Here's how a typical robotic work cell design simulation would work:

1. Before simulating a cell, a geometric model of the robot, gripper, and attendant machinery needed on a graphic display is built.
2. Each component, now modeled in 3D, is brought to the screen and graphically placed in the work cell.
3. The task is developed and then animated; motion is illustrated and problem areas identified.

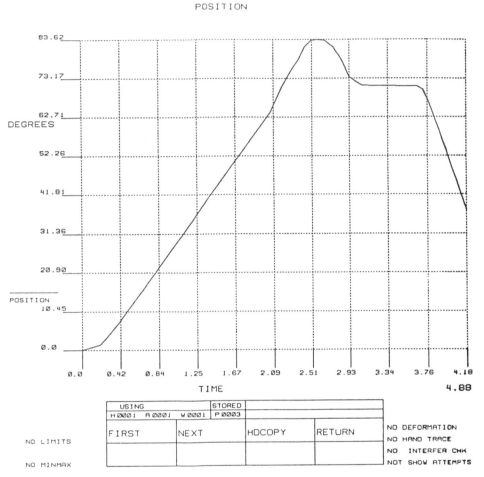

POSITION

DEGREES

POSITION

TIME

USING			STORED	
H 0001	A 0001	W 0001	P 0003	
FIRST		NEXT	HDCOPY	RETURN

NO LIMITS

NO MINMAX

NO DEFORMATION
NO HAND TRACE
NO   INTERFER CHK
NOT SHOW ATTEMPTS

Fig. 10.4   Graph of position vs. time for the first joint of the robot of Figure 10.3 as it moves from the work table to conveyor.

10.10 HOW TO SIMULATE AN AUTOMATED QUALITY ASSURANCE PROJECT

In selecting a specific simulation package, the user should go beyond looking at computer compatibility and modeling capabilities and consider how easy the software will be to use and how applicable it is to the specific tasks for which it will be used. Contacting someone experienced in the use of the simulation packages can be invaluable.

General Electric Company is one of the biggest users of simulation for robotics and manufacturing in the United States. Based on the experience of performing simulation for plants around the country, they have developed the following eleven steps that have proven to result in accurate models of flexible manufacturing systems:

1. Define the goal of the project so the simulation can be tailored to meet the goal.
2. Make assumptions on the equipment to be installed.
3. Build the software model of the actual facility and equipment.
4. Gather data on the various known parameters of the physical system being simulated.
5. Develop the program using the most appropriate computer language.
6. Verify the program by running it under simple assumptions for which the results are known or easily computed.
7. Validate the program using goodness-of-fit tests to check that the computer program's output resembles output data taken from a real system.
8. Test the model under varying conditions and assumptions usually at extremes, to check on the sensitivity at both ends of the model.
9. Analyze the output data, all event times and changes in the simulation project by means of statistical and graphical techniques.
10. Record the results of the simulation to aid management decision making and successful implementation of the system.
11. Train the user of the system so that the software used for simulation purposes can also be hooked up for actual scheduling systems used for factory production.

10.11 ASSESSING THE BENEFITS

To perform a cost-benefit analysis, the benefits of automated quality assurance can be quantified. Tangible benefits include:

Lower inventory
Reduced labor costs
Less scrap and rework
Fewer engineering prototypes
Reduced changeover costs and time
Lower energy and raw material consumption
Eliminated data re-entry and processing duplication

The imagination comes in pricing intangibles like:

Competitive standing
Better quality control
Faster product introduction
Consistent and tireless operation
Improved on-time product delivery
Using standard component designs and processes
Reduced design and manufacturing lead time
Increased flexibility in design and production

Of the 700,000 different quality control tests run regularly in the United States alone, it is estimated that at least 25 percent could be replaced by fully automated machine vision inspections. An additional 40 percent could be more effectively handled by an operator using machine vision as a gauge.

Vision systems purchased for stand-alone inspection applications generally have good paybacks. Relatively low system costs and simple work cells can easily be balanced by direct labor savings. When the vision system is purchased as a component of a complex manufacturing work cell, the system cost would generally be only a small portion of the installation cost.

The follolwing issues have relevance to payback justification of a vision system beyond direct labor or material savings:

What is the value associated with increased quality control or 100 percent inspection?
Can inspection information be used as feedback for process control to improve quality?
Would robot perception open possibilities for automation not otherwise feasible?

Can inspection speeds associated with automation allow faster operation of a production line?

What cost savings are associated with decreased scrap or reject rates?

Is a vision system the only automation technology for a certain type of inspection or process?

Can a vision system provide "flexible software fixturing" to eliminate conventional fixturing which is replaced during new model retooling?

11

Artificial Intelligence: The Next Step

11.1 Expert and Knowledge-Based Systems
11.2 Expert Systems for Quality Assurance
11.3 The Technology of Expert Systems
11.4 Tools for Building Expert Systems
Bibliography

11.1 EXPERT AND KNOWLEDGE-BASED SYSTEMS

Expert systems are currently the most emphasized area in the field of artificial intelligence (AI) and represent the leading edge of commercialization in computer science. Professor E. Feigenbaum of Stanford University, a pioneer in the field of artificial intelligence, defines an expert system as: an intelligent computer program that uses knowledge and inference procedures to solve problems that are difficult enough to require significant human expertise for their solution. The knowledge necessary to perform at such a level, plus the inference procedures used, can be thought of as a model of the expertise of the best practitioners of the field.

The knowledge of an expert system consists of facts and heuristics. The facts constitute a body of information that is widely shared, publicly available, and generally agreed upon by experts in a field. The heuristics are mostly private, little-discussed rules of good judgment (rules of plausible reasoning and good guessing) that characterize expert-level decision making in the field. The performance level of an expert system is primarily a function of the size and quality of the knowledge base that it possesses.

The potential uses of expert systems appear to be virtually limitless. They can be used to diagnose, monitor, analyze, interpret, consult, plan, design, instruct, explain, learn, and conceptualize. Thus, they are applicable to mission planning, monitoring, tracking and control, communication, signal analysis, command and control, intelligence analysis, targeting, construction and manufacturing (design, planning, scheduling, control), education (instruction, testing, diagnosis), equipment (design, monitoring, diagnosis, maintenance, repair, operation, instruction), image analysis and interpretation, professions (law, medicine, engineering, accounting, law enforcement), consulting, instruction interpretation, analysis, software (specification, design, verification, maintenance, instruction), and weapon systems (target identification, electronic warfare, adaptive control).

An expert system consists of:

A knowledge base (or knowledge source) of domain facts and heuristics associated with the problem

An inference procedure (or control structure) for utilizing the knowledge base in the solution of the problem

A working memory—"global data base"—for keeping track of the problem status, the input data for the particular problem, and the relevant history of what has been done

A human "domain expert" usually collaborates to help develop the knowledge base. Once the system has been developed, in addition to solving problems, it can also be used to help instruct others in developing their own expertise.

Artificial intelligence researcher Michie observes that [ideally] there are three different user modes for an expert system in contrast to the single mode (getting answers to problems) characteristic of the more familiar types of computing: (1) getting answers to problems with the user as client; (2) improving or increasing the system's knowledge with the user as tutor; and (3) harvesting the knowledge base for human use with the user as pupil. Users of an expert system in mode (Davis and Lenat, 1982) are known as "domain specialists." It is not possible to build an expert system without one. An expert system acts as systematizing repository over time of the knowledge accumulated by many specialists of diverse experience. Hence, it can and does ultimately attain a level of consultant expertise exceeding that of any one of its "tutors."

It is desirable, though not yet common, to have a natural language interface to facilitate the use of the system in all three modes. In some sophisticated systems, an explanation module is also included, allowing the user to challenge and examine the reasoning process underlying the system's answers. When the domain knowledge is stored as production rules, the knowledge base is often referred to as the "rule base," and the inference engine as the "rule interpreter."

An expert system differs from more conventional computer programs in several important respects. Duda observes that in an expert system, ". . . there is a clear separation of general knowledge about the problem (the rules forming a knowledge base) and methods for applying the general knowledge to the problem (the rule interpreter)." In a conventional computer program, knowledge pertinent to the problem and methods for utilizing the knowledge are all intermixed, making it difficult to change the program. In an expert system, ". . . the program itself is only an interpreter (or general reasoning mechanism) and [ideally] the system can be changed by simply adding or subtracting rules in the knowledge base."

The development of any program in artificial intelligence involves a great deal of computer programming. This is especially true when a program is started without reference to other programs so that everything must be worked out on a completely new basis. In developing various types of expert consultation programs, it was realized that much of the methodology (and possibly some of the programming) could be used to develop expert programs in areas outside the original program, or knowledge engineering tool.

What is knowledge engineering? In simplest terms, it is the codification of a specific domain of knowledge into a computer program that can solve problems in that domain. The task involves the cooperation of human experts in the domain working with the program designer and/or knowledge engineer to codify and make explicit the rules that a human expert uses to solve real problems. The expert often uses rules applied almost subconsciously, and the program usually develops in what may seem to be a hit-or-miss method. As the rules are refined by using the emerging program, the expertise of the system increases. As the knowledge of more and more human experts is incorporated in the program, the level of expertise rises and eventually can exceed that of any specific human expert.

Knowledge engineering usually has a synergistic effect. The knowledge possessed by human experts is often unstructured and not explicitly expressed. The construction of a program aids the expert in learning what is known, and at the same time, can pinpoint inconsistencies between one expert and another. Major goals in knowledge engineering include the construction of programs that are modular in nature, so that additions and changes can be made to one module without affecting the workings of other modules. A second major goal is the obtaining of a program that can explain why it did what it did and when it did it. If the program evokes rule 86 to explain a certain set of facts, and if the human expert questions the correctness of applying this rule to the data, the program should be able to explain why it used rule 86 instead of, say, rule 89. This type of interaction allows rules to be refined and brings to light inconsistencies in procedures and data. Feigenbaum defines the activity of knowledge engineering as follows:

> The knowledge engineer practices the art of bringing the principles and tools of AI research to bear on difficult application problems requiring experts' knowledge for their solution. The technical issues of acquiring this knowledge, representing it, and using it appropriately to construct and

explain lines-of-reasoning, are important problems in the design of knowledge-based systems. The art of constructing intelligent agents is both part of and an extension of, the programming art. It is the art of building complex computer programs that represent and reason with knowledge of the world.

11.2 EXPERT SYSTEMS FOR QUALITY ASSURANCE

Specifications for manufactured products are influenced by a variety of factors, including governmental regulations, liability, consumer requirements, and economics. These factors often change, resulting in a modification of quality assurance requirements. An expert system can serve to assist a company in assessing and keeping up with such changing conditions. When corporate policies on quality control change, for example as the result of a new federal regulation, an expert system overlaying the manufacturing process could implement that change wherever appropriate in production and inspection procedures.

Industry is beginning to move from a defect-detective approach to a defect-preventive approach. Future machine controllers will incorporate expert systems which modify manufacturing variables to improve the operation before it even begins.

While the actual use of expert systems for quality assurance is only beginning, some example applications have been reported. Arthur D. Little, Inc. has developed Product Labels and Product Safety, two expert systems for quality assurance applications. Product Labels assists in the preparation of a legally acceptable product label for consumer and industrial products. It assesses the relative risk of using different labeling schemes, under a variety of conditions. Product Safety helps manage and identify the myriad of risks associated with bringing a new product to market. It focuses on industrial chemicals and similar commercial products.

Teknowledge, Inc., a California-based AI firm, reports the development of Component Evaluation, an expert system that evaluates proposed changes in a components mix for quality control. It was built using Teknowledge's M.1 on a PC AT clone. Quality Control Advisor is another expert system developed using Teknowledge's M.1. Its application is in the area of batch manufacturing.

The State College, PA, plant of Corning Glass has developed an expert system to diagnose breakage problems that occur at the lehr. Product defects caused by other processing steps manifest themselves when the ware cools as it leaves the lehr. Diagnosing breakage helps isolate the upstream problem that requires correction. The system was developed using Texas Instrument's Personal Consultant.

Beckman Instruments, Inc., has developed SPINPRO, an expert system that designs ultracentrifugation procedures. The investigator and the program participate in a question and answer dialogue in which the research goals and sample characteristics are defined. This interaction with the program is modeled on the way customers currently interact with human experts by telephone. At the conclusion of the dialogue, SPINPRO produces four reports. One report summarizes the questions posed by SPINPRO and the answers provided by the investigator. The second report describes an optimal ultracentrifugation procedure to achieve the stated goals using the optimal equipment. The third report is similar, but outlines a procedure based upon, and constrained by, the centrifuge equipment available in the investigator's laboratory. A fourth report compares the two procedures and the effectiveness of each in performing the run. Thus, the program performs the role of an expert advisor, offering knowledgeable advice and comparing alternatives. SPINPRO has shown that it can reduce run times and improve the quality of separations. The investigator can more easily design procedures precisely fitted to the particular requirements of the research. Supporting information and calculations are readily available. SPINPRO addresses the problem of improving the performance of the ultracentrifugation laboratory by working with the investigator to design efficient ultracentrifugation procedures.

Pharmaceutical companies are reportedly experimenting with expert systems in testing of drugs for side effects. The expert system would provide rules to overlay a database, alerting the company to potential problems and advising on what should be reported to the FDA. Many medical expert systems involve drug usage—this application is a spinoff of this type of expert system.

There is a considerable effort to apply expert systems to robotic software systems and languages, including graphics and simulation techniques. In the Robotics Software Project development program at CAM-I, prototype software is being developed within the framework of the three subsystems. It is reported that expert systems are being developed for the control system subsystem. Research at Vanderbilt University is being initiated by Professor Kazuhiko Kawamura to develop expert systems using PROLOG as a kernel language. Future applications include management of complex systems and planning and control of intelligent robots. In a research project at Georgia Tech, Dr. William Underwood is investigating the knowledge and reasoning required to automatically produce robot plans to fabricate aircraft parts.

The Jet Propulsion Laboratory is working on a three-part, closed-loop control system that is directly applicable to robots. The system has three parts: (1) DEVISER, a planner/scheduler expert system; (2) FAITH, a systems diagnostician; and (3) PEER (Planning and Execution with Error Recovery), to link the other two modules together. The control is a predicate logic rule-based system. The system would operate in a robot as follows: Goals are entered into the planner/scheduler which constructs a plan and comes up with a set of measurable sensor states. The execution monitor detects errors by comparing the predicated states with the actual measurements it makes with the sensors. Any discrepancy between the two indicates that something is wrong. The diagnostician then determines what actions should be taken to recover from the error (contact: Leonard Friedman)

Another expert system development is at Purdue University, Department of Electrical Engineering, under the direction of Professor R.L. Kashyap. ROPES (Robot Planning Expert System) is a robot planning system being built up using the concept and technology of the expert system and other ideas of artificial intelligence. This project uses the Prolog (C-Prolog) programming language to construct the system prototypes which include several different cases. The simulations of the models have been finished by testing them in the Purdue Engineering Computer Network (PECN) which utilizes the UNIX operating system. The simulation results have been compared with the other robot planning systems and this robot planning system is much more effective than the others.

McDonnell-Douglas has three projects underway using expert systems to solve problems in robotic off-line programming. One project involves artificial intelligence methods in collision detection and avoidance, as well as in robotic grasp and trajectory planning. Another project deals with off-line robot dynamics and involves sensory reasoning. The third expert system project will guide a robot in deciding what to do next in a cell of six to eight machine tools.

Adaptive Technologies, Inc., was one of the first companies to offer a commercial expert system for robotics. Their expert system, ADAPTIWELD, is used in conjunction with robotic arc welding systems and incorporates knowledge of skilled welders in its information and control base. A three-dimensional vision system gathers information on the characteristics of a seam to be welded. These characteristics are stored in the computer memory of the system controller and manipulated by the expert system to perform a complete welding function without direct human supervision. The expert system approach allows the system to perform autonomous welds, and the user easily adapts to the system for unique welding requirements. The knowledge base allows the operator to transfer knowledge to the welding system.

Electrical Laboratory has developed SPIDER, an expert system for image processing. Because it has about 300 sub-routines, SPIDER is not easy for a non-expert on image processing to select and combine appropriate sub-routines for image processing purposes. DIA-EXPERT is an expert system advisor for SPIDER and is used to synthesize desired image processing. The system has two main parts. A semantic part to analyze the objective of the user and select an appropriate set of sub-routines, and a structural part to check the structural consistency of the selected sub-routine and generate a series of execution programs.

There is an almost universal agreement among machine vision experts that artificial intelligence will be a major factor in the growth of the field—the only question is when. In an interview with expert Systems Report, Dr. Linda Shapiro, director of intelligent systems at Machine Vision International, expressed the view that artificial intelligence for machine vision isn't far off. The next generation of machine vision systems will have to be able to reason; that is the only way these systems will be able to carry out the necessary high-level processing.

Researchers at Stanford University are developing ACRONYM, a model-based image understanding system. It demonstrates mechanisms for the interpretation of images with generic object classes and generic viewing conditions, in a way that is generalizable. It incorporates a powerful geometric modeling capability with a high level modeling language for natural communication with the user in terms of object models. A user gives high level descriptions of both generic and specific instances of objects. A rule-based inference system produces a viewpoint dependent symbolic summary of the predicted appearance of the objects. This geometric reasoning capability enables the system to incorporate and relate knowledge and information at different levels. This summary drives a powerful syntactic matcher to find instances of the objects in the preprocessed images (Stanford University).

IMAGEX, an ongoing research project at UCLA, is an image oriented expert system that allows the selection of pictorial images for inclusion in reports (such as a corporate annual report). IMAGEX can help guide the use of portraying one particular visual "image" for a set of data or contextual situation, and can help avoid selection of an image with improper connotations. Future uses of the expert system are found in the communication and advertising realms.

DIPS is an image processing system offered commercially by Dalek Corporation. It is written in LISP and Flavors for Symbolics 3600 series machines. DIPS includes commands for image manipulation, edge detection, texture filtering, histogramming and thresholding, arithmetic and boolean operations, and local and global image segmentation.

11.3 THE TECHNOLOGY OF EXPERT SYSTEMS

While the quality control engineer's involvement with may involve only the use of off-the-shelf expert systems, an understanding of how this software works is still important. This section is designed to provide an insight into expert system technology.

Artificial Intelligence Programming Languages

There are three general families of languages used for artificial intelligence programming: (1) functional applications languages such as LISP, (2) logic programming languages such as PROLOG, and (3) object-oriented languages such as SMALLTALK and ACTOR.

Many artificial intelligence (AI) programs are written in LISP (LIST Processing), which was developed by John McCarthy at MIT in the early 1960s. LISP is an easy programming language to learn, primarily because it has a simple syntax. LISP programs and data are both in the same form-lists. Thus, AI programs can manipulate other AI programs. This allows programs to create or modify other programs, an important feature in intelligent applications. It also allows programming aids for debugging and editing to be written in LISP, providing great interactive flexibility for the LISP programmer.

PROLOG is a theorem-proving, logic-oriented language developed in 1973 at the University of Marseille AI Laboratory by A. Colmerauer and P. Roussel. Additional work on PROLOG has been done at the University of Edinburgh in Great Britain. Development of PROLOG in France has achieved a documented system that can be run on nearly all computers. PROLOG is very popular in Europe and is targeted as the language of Japan's fifth generation computer project. PROLOG's design (and its powerful pattern matcher) is well suited to parallel search and is therefore an excellent candidate for future computers that incorporate parallel processing. Substantial interest in PROLOG is now arising in the United States, with some of PROLOG's features being implemented in LISP.

An emerging trend is the increasing use of object-oriented programming to ease the creation of large exploratory programs. The use of objects, a good way to program dynamic symbolic simulations, will become more important as both the quest for utilizing deeper knowledge and the demand for increased reliability of knowledge-based systems accelerates. Object-oriented programming also holds promise for distributed processing, as each object could be implemented on a separate processor in a linked network of processors.

There is open controversy regarding which language is best for artificial intelligence and multiprocessor programming. LISP, a functional programming language, has always been most popular in the United States. PROLOG, a logic programming language, was chosen for the Japanese fifth generation project and is gaining some support in North America. Only the next few years will determine which language becomes dominant or if both will remain equally popular in different programming circles.

The programming environment of LISP has evolved over two decades. Consequently, it is substantially easier to develop a powerful AI program in LISP than PROLOG as the programming environment PROLOG offers is relatively poor. This drawback, however, is due to the lack of a long history of use, and may be overcome in the future.

Knowledge Representation

While the representation of knowledge in the form of production rules is the most common technique, not all expert systems are rule-based. MACSYMA, INTERNIST/CADUCEDUS, DIGITALIS THERAPY ADVISOR, and PROSPECTOR are common examples of network-based expert systems. Frames are also a popular method of representing knowledge, and most of the powerful commercial tools for building expert systems employ a combination of frame and rule representation.

Production Systems

Production systems solve problems by searching through the space of possible problem states (e.g., the state space) that correspond to the condition of the problem at each stage of its solution. In chess, for example, the initial configuration of the playing board represents a problem state from which several alternative states can be generated, depending on the opening move. Each of these new states can be produced by the application of one of the rules for moving chessmen. The object is to modify the problem state progressively, from its initial state to a goal state (i.e., checkmate), by the application of an appropriate sequence of rules.

Production rules have proven to be such a convenient modular way to represent knowledge, they now form the basis of most expert systems.

Production Rules

In a production system, the rules (e.g., productions) are inference rules that are based upon what is known about the problem in general. For example, it is known that all mammals have hair. If it is known that an animal has no hair, then it can be inferred that it is not a mammal. This simple deduction amounts to a change in the problem state if there is concern with identifying animal species. That is, what is known about the problem in particular (the data base) now includes [not mammal], which narrows the focus of the search and advances search progress beyond the initial state. Each of the rules consists of a condition part and an action part, and has the following general form:

IF: [antecedent]

•

•

•

[antecedent]

THEN: [consequent]

with certainty C

•

•

•

[consequent]

with certainty C

If all the antecedents can be matched against assertions in the global data base (GDB), then the consequents can be performed. The rule IF: ([has no hair] THEN: [is not mammal] specifies a condition (has no hair) that must be present in the global data base before the production can fire (i.e., before the action specified by the consequent can be executed).

In more complex systems, a very complex control structure may be used to decide which group of production rules to examine, and which to execute from the production rules in the group that match patterns in the global data base. In general, these control structures work in a repetitive cycle of the form:

1. Find the "conflict set" (the set of competing rules which match some data in the GDB).
2. Choose a rule from among the conflict set.
3. Execute the rule, modifying the GDB.

Production rule systems can be implemented for any of the problem-solving approaches discussed earlier. Thus, the "top down" approach may be used, employing the rules to chain backwards from the goal to search for a complete supportive or causal set of rules and data ("goal-driven," or "model-driven" control structure). Or, a "bottom-up" approach can be used, employing forward-chaining of rules to search for the goal ("event-driven" or "data-driven" control structure).

In complex systems employing many rules, the control structure may contain meta-rules which select the relevant rules from the entire set of rules, and also focuses attention on the relevant part of the data base. This reduces the search space to be considered. The control structure then employs further heuristics to select the most appropriate rule from the conflicting rules which match the preconditions in the global data base.

The Rule Interpreter and System Operation

The function of the rule interpreter is to examine the production rules to determine which ones are enabled (i.e., capable of being fired) and, after resolving which rule to apply, to fire the selected production. The control strategy of the rule interpreter determines how the enabled rules are found and where to apply them. Forward-chaining is one of the simplest strategies and consists of scanning the rules for matching antecedents, applying the rule found and updating the data base. The actions of the fired productions alter the contents of the data base by changing an assertion or by adding a new one (e.g., animal is not mammal) such that other rules may become enabled or disabled. Thus, when a rule is applied, an inference is made and registered in the global data base, thereby generating a new problem state from which the search may be advanced. The process continues until a goal state is reached or no applicable rules are found. This was the strategy followed in the chess example, where the problem state was brought forward from its initial configuration to a check mate configuration by the application of an appropriate sequence of rules. Backward-chaining consists of selecting a goal and scanning the rules to find those whose consequent actions can achieve that goal.

Semantic Networks

Certain types of relationships can be made clear best by using a graphical presentation. Just as mathematical formulas can often be clarified by being presented graphically, so can complex ideas. Similarly, the computer programs are often difficult or impossible to understand when written as lines of code, even in a high level language, but can be understood when presented as a flow chart.

A semantic network is an approach to describing the properties and relations of objects, events, concepts, situations or actions by a directed graph consisting of nodes and labeled edges (arcs connecting nodes). Because of their naturalness, semantic networks are very popular in artificial intelligence.

Frames

One of the key ideas for knowledge representation is the use of frames, where a frame is a collection of facts and data about some thing or some concept. Most of the powerful tools for building expert systems allow knowledge representation by both production rules and frames. The concept of frames was developed by Marvin Minsky of MIT.

A frame of schema representation is based on the theory that previous situational experiences create certain expectations about objects and events associated with new situations, and provides a framework within which new information can be interpreted. That is, a frame is a structure within which data or knowledge about stereotyped situations can be represented. For example, based on previous experience, a chair is generally expected to be a kind of furniture with arms, legs, and a back. The expectations represent things that are always true about chairs and provide the context within which other objects can be interpreted. These expectations are represented as terminals or slots within the framework or context of the situation. The slot provides a mechanism for a kind of reasoning called expectation-driven processing. Empty slots (i.e., unconfirmed expectations) can be filled, subject to certain assignment conditions, with data that confirm the expectations. Thus, frame-based reasoning is based on looking for confirmation of expectations and is often just filling in slot values. One of the important ways in which slot values are specified is by inheritance.

Perhaps the simplest way to specify slot values is by default. The default value is attached loosely to the slot so as to be easily displaced by a value that meets the assignment condition. In the absence of information, however, the default value remains attached and expressed.

Finally, procedures can be attached to slots and used to derive slot values. So-called slot-specific heuristics are domain-specific procedures for deriving slot values in a particular context. An important aspect of attached procedures is that they can be used to direct the reasoning process. In addition to filling in slots, they can be triggered when a slot is filled, possibly altering or terminating the slot filling process.

An example of a frame is:

Airplane Frame:
Type:
      range: (fighter, transport, trainer, bomber, light plane, observation)

Manufacturer:
    range: (McDonnell-Douglas, Boeing . . .)
Empty Weight:
    range: (500 lbs to 250,000 lbs)
Gross Weight:
    range: (500 lbs to 500,000 lbs)
    if needed: (1.6 x empty weight)
Max Cruising Range:
    if needed: (look up in table cruising range appropriate to
            type and gross weight)
Number of Cockpit Crew:
    range: (1 to 3)
    default: 2

Attached to each frame is information on how to use the frame, what to do if something unexpected happens, and default values for slots.

Frames can also include procedural as well as declarative information. Frames facilitate expectation-driven processing—reasoning based on seeking confirmation of expectations by filling in the slots. Frames organize knowledge in a way that directs attention and facilitates recall and inference.

Search

The term search always refers to the search for a goal achieving sequence.

The problem of finding a route from an initial state or combination of elements to a desired state involves a search and examination of intermediate states. Any specific intermediate state may be closer to the desired state than all the others; it may be no closer or further away than any other; it may be on a path that can or must terminate in the desired state; it may be on a path that will never reach the desired state; it may be on a path that will never reach the desired state by an indirect or inefficient route rather than by a best route, etc.

In complex problems, exhaustive search is often impossible because it results in what is known as a combinational explosion. Some of the most startling examples of combinational explosions occur in games such as chess and checkers. If, at the opening move of a checkers game, a player attempts to examine every possible move both players could make until the game was finally won or lost, based on the first move, approximately ten configurations would have to be examined. If the examinations could be performed at the rate of three billion per second, the process would take about 10 centuries. Chess is even more complex, with about ten possible configurations.

No computer program, whether for AI or other purposes, can afford to get trapped in a combinational explosion. Thus, a number of search strategies have been developed to overcome this problem (when possible). One of the most powerful of the search strategies is the use of heuristics. Heuristics is the application of what might be called common-sense or domain specific knowledge to the problem at hand.

Another important search technique involves the calculation and assignments of numerical values to each state that is examined. These numerical values attempt to rate the state as being likely or unlikely to lead to a solution. In this way, the most likely states are examined first.

Other search techniques include:

1. The Alpha-Beta procedure. This technique calculates upper and lower numerical bounds on each state (or node) and can reduce search by orders of magnitude.

2. Hill climbing. In this procedure, search proceeds in whatever direction seems to be most promising. Sometimes the hill turns out to be only a local perturbation and the strategy then fails.

3. Branch and bound. This technique finds the shortest path to a goal state; it works by assigning cost to each partial path.

4. Best first. Generally finds a more optimal path to a goal than breadth-first or depth-first. (Note that only one path if found in breadth-first or depth-first search because once a path is found, the search stops. This path may or may not be optimum.)

5. AND/OR trees. Many problems can be solved most easily by breaking them up into smaller problems.

6. Minimaxing.   This technique is used in games. One evaluates a possible move in terms of the maximum advantage it will confer compared to the minimum advantage. Choosing the move with minimum advantage guarantees some forward progress. Opting for maximum advantage usually means that one will be forced by the opponent into a position worse than that associated with the minimum.

Constraints

In searching for a solution to a problem, blind alleys, useless paths, and combinatorial explosions should be avoided if possible. One way to reduce the amount of search to be used in solving a problem is to find characteristics that can be used to increase the efficiency of the search. The use of constraints has been found useful in all artificial intelligence programs.

During scene analysis in the world of toy blocks, for example, straight lines may meet for form a L-shape, an arrow, a T, an X, and several other figures. A line may represent an outer edge of a block, or a boundary between a block and the background, or an interior edge, etc. Conventions have been developed for labeling these lines and figures to indicate its nature and how it is to be handled by the program. A single L, for example, could theoretically have as many as 2500 different labels; when the number of physically possible labelings are considered, it equals only 3.5 percent of the total. A certain type of form peak has 6,250,000 (approximate theoretically possible labels, but the actual number of physically possible labels is only 10, or 0.0016 percent of the total. The use of constraints in the form of physically possible labels in this situation has reduced an incredible, complicated data processing problem to one much more manageable.

In sentence parsing, similar concepts about word position, which words are nouns, verbs, or prepositions, and the kinds of things nouns do and have done to them by other words makes parsing by machine possible.

Backtracking

When a program moves from one state to another, it may find that the new state takes it farther away from its target goal state than it was before it made the move. In this case, the program may decide to return to a previous state. The process of going back to a state that has already been tried is called backtracking. If a program gets involved in too much backtracking, it spends too much effort in useless explorations and thus program efficiency decreases. As a result, most programs try to reduce backtracking as much as possible or even eliminate it entirely whenever it is possible to do so.

Forward and Backward Chaining

Typically, an artificial intelligence program will have an initial state and a desired goal state. To get from the initial state to the goal state normally involves passage through a long chain of intermediate steps or states. When the program works from the initial state toward the goal state, the process is called forward chaining. Forward chaining is a good technique to use when all or most paths from any one of many initial or intermediate state converges on one or a few goal states. In this case, the program begins to look for a path through the problem by starting at a goal state and seeing how it can be modified to bring it closer to an initial state. Backward chaining is an efficient technique to use when any one of many goal states will satisfy the requirements of the problem while the initial states are few in number; the situation in this case represents many goal states converging on one or a few initial states.

When data or basic ideas are a starting point, forward chaining is a natural direction for problem solving. It has been used in expert systems for data analysis, design, diagnosis, and concept formation. Backward chaining is applicable when a goal or a hypotheses is a starting point. Expert system examples include those used for diagnosis and planning.

When the search space is large, one approach is to search from both the initial state and the goal or hypothesis state and utilize a relaxation type approach to match the solutions at an intermediate point. This approach is also useful when the search space can be divided hierarchically, so that a bottom-up and top-down search can be appropriately combined. Such a combined search is particularly applicable to complex problems incorporating uncertainties, such as speech understanding.

11.4 TOOLS FOR BUILDING EXPERT SYSTEMS

The time required to develop the early expert systems, often ten to twenty man years, was a big drawback to those considering new applications. A large portion of the development time was consumed in developing knowledge representation schemes and reasoning strategies that may be applied to problems in numerous domains. Artificial intelligence researchers began to develop domain-independent expert systems, commonly called "tools for building expert systems." By using such tools, the knowledge and engineer's task is simplified to eliciting knowledge from an expert and representing this knowledge in a form suitable for the expert system. The laborious task of building the inference engine is eliminated. In the 1970s and early 1980s, about a dozen such expert system development tools were built by universities and other research groups. Among the most popular of these systems were EMYCIN, OPS5, and ROSIE. In 1983, several AI firms announced the availability of expert system development programs. In early 1984, some of these tools began to become commercially available. These commercial systems were easier to use and better supported than the academic systems. The marketing of these programs has great business potential.

BIBLIOGRAPHY

Barr, A. and Feigenbaum, E.A. (1981, 1982). *The Handbook of Artificial Intelligence*, Vols I, II, Los Altos, CA, W. Kaufman.

Davis, R. and Lenat, D.B. (1982). *Knowledge Based Systems in Artificial Intelligence*, New York, McGraw-Hill.

Gevarter, W.B. (1983). *An Overview of Artificial Intelligence and Robotics*, National Bureau of Standards, NTIS # N83-31379.

Hayes-Roth, F. (ed.) (1983). *Building Expert Systems*, Reading, MA, Addison-Wesley.

Miller, R.K. (1986). *Computers for Artificial Intelligence*, SEAI Technical Publications, Madison, GA.

Rich, E. (1983). *Artificial Intelligence*, New York, McGraw-Hill.

Walker, T.C. and Miller, R.K. (1986). *Expert Systems 1986*, SEAI Technical Publications, Madison, GA.

12

Case Studies

12.1 Allen-Bradley
12.2 AT&T
12.3 Deere & Company
12.4 Frost, Inc.
12.5 General Electric
12.6 General Motors
12.7 Merck & Company, Inc.
12.8 Texas Instruments
12.9 3M
12.10 Whirlpool

Ten leading U.S. corporations involved in computer-integrated manufacturing were selected for analysis. The sections which follow demonstrate corporate strategies aimed at both insuring product quality
and implementing computer-based manufacturing technologies to
advance corporate productivity.

12.1 ALLEN-BRADLEY

Allen-Bradley, a subsidiary of Rockwell International, is a manufacturer of controls for automation, and offers communications
expertise and equipment needed for users to create a fully integrated
automation system. The firm demonstrated computer-integrated
manufacturing capabilities in-house with its Milwaukee plant.

The Total Quality Management
Systems Program

In a speech at the Autofact '87 conference, Allied Bradley Chairman C.R. Whitney described how the company embarked on a re-
positioning strategy seven years ago. A basic objective of the plan
was to become a world class manufacturer.

Allen-Bradley had always had a proud tradition of quality. In
fact, the word "quality" has been in the corporate logo for nearly
50 years. But, the Allen-Bradley quality program was only an in-
spection-based system. Dr. Deming and Mr. Juran told the company that that wouldn't "cut it" in the world marketplace. The
new company approach to quality was directed at a prevention-
based system. Subsequently, management learned that this effort
had to be broadened to include all aspects of the business, not just
manufacturing. thus evolved the Allen-Bradley Total Quality Management Systems (TQMS), a system which commands the attention
of every employee, from top to bottom. TQMS turned the whole
company into one huge quality assurance machine, helping them
control quality throughout a product's life cycle.

Total Quality Management System made quality a strategic
objective—as important as financial performance—and it gave the

firm the means to measure quality as precisely as they measure financial performance.

The reason TQMS works is because it has the full participation of company management. What does management do? For one thing, it sets quality improvement goals for each major business group. And it requires each division to routinely report progress with prevention, appraisal, scrap, rework, warranty returns, supplier quality, and customer satisfaction.

Management makes TQMS an important part of management performance reviews, with quality objectives incorporated into job descriptions and quality improvements tied directly to compensation packages. This management commitment gives Allen-Bradley people the backing they need to focus the entire organization's efforts on the persuit of zero defects.

The program has paid off handsomely. In its first five years, Whitney estimates that it saved Allen-Bradley $82 million in quality costs. TQMS lowered the total cost of quality for the leading product line—programmable controllers—by nearly two-thirds, from 13 percent of sales to less than five percent. On a cumulative basis, management estimates that for every dollar invested in the TQMS program, the company got 13 in return.

TQMS also helped improve product reliability. It took programmable controllers from hundreds of hours of failure-free operation to hundreds of thousands of hours. TQMS helped trim warranty returns and warranty costs, reducing them by nearly 25 percent over a three-year period. Customers noticed improvements and purchased more products, helping Allen-Bradley to more than double their programmable controller market share.

The Milwaukee Plant

Allen-Bradley's new electrical contactor facility in Milwaukee is a $15 million, 50-machine flexible-assembly complex that reads parts-insertion requirements from bar codes and can produce motor starters in 125 different configurations at the rate of 600 an hour. Manufacturing, assembly, shipping, and packaging operations are integrated.

The 45,000 square-foot facility is located on the eighth floor of Allen-Bradley's main plant. The installation also serves as a showcase of the company's latest automation control technology. Integration engineering for the system was done by the company's Industrial Automation Systems Division. The facility has two doors. Materials such as brass, steel, silver, molding powder, coils and springs enter through one door and the finished products exit through the other. In one corner, three plastic molding machines produce housings for the contactors; in the opposite corner, twelve machines produce components such as terminals and spanners. In the only manual handling operations in the facility, housings and parts are moved in cartridges to the assembly line. In all, 26 machines, including subassembly, assembly, testing and packaging, are used to manufacture two sizes of motor starters in 125 variations. More than 60 percent of the machinery was designed and built by Allen-Bradley personnel. One of the first subassembly machines puts a printed bar code on the base of the unit. By scanning the bar code, assembly cells are automatically instructed to perform specific tasks. Photosensors direct the flow of product through the assembly cells and staging lines.

The continuous flow of the computer integrated manufacturing (CIM) line provides the company with a virtual zero-inventory status. The product is marshalled and on its way to the shipping department in less than five days, compared to a previous contactor cycle of 142 days "wall-to-wall." The value contained in that earlier inventory structure is approximately equal to the $15 million the company invested in the CIM project.

According to Allen-Bradley, tangible benefits include:

Reduced labor (both direct and indirect), and therefore less supervision

Lower inventory costs, due to elimination of work-in process and finished goods inventories

Reduced costs, due to higher quality

Reduced energy consumption, due to more efficient operations

Intangible benefits are:

More consistent product quality, because direct labor is eliminated

Fewer future capital expenditures, because future quality is in
place now

Future flexibility, due to ability to increase product variations
as markets dictate

Reduced downtime, more productive uptime, due to advanced
diagnostics

Increased inventory savings and customer satisfaction, due to
faster turnaround

The performance of the contactor plant has contributed significantly to the company's export growth—from 5 percent of total
sales in 1979 to 5 percent in 1985. One measure of Allen-Bradley's
satisfaction with its automation strategy: The company is currently
building new state-of-the-art electronics facilities in Milwaukee and
Great Britain and adding flexible automation at an existing sheet-
metal operation.

12.2 AT&T

The AT&T Richmond, Virginia, printed circuit board facility is
one of the country's most advanced manufacturing plants. In 1984,
the facility won the Society of Manufacturing Engineers LEAD
Award, with the citation reading:

> The fully-integrated manufacturing control system is used
> in the automation of printed wiring boards. The compu-
> terized monitoring system permits close control of complex
> manufacturing operations and has brought significant sav-
> ings in cost and production time.

The benefits of the CIM program at the Richmond works are
numerous. The plant has experienced a reduction in both tooling
manufacturing intervals which results in better customer service.
Direct labor cost has been reduced, resulting in lower product cost.
There have been benefits in quality assurance. An intangible benefit

is the human factor. The CIM system allowed employees to be freed from mundane tasks and allowed them to be more versatile employees. Because of increased satisfaction and self-esteem, the user has become an integral part of the total system instead of just a button pusher. Other benefits derived from the system include manufacturing flexibility and simplified rate calculation, cost estimation, and capacity planning, all made possible by having all data stored on a common database.

Some statistics which indicate the support offered by the AT&T CIM system are:

100,000 MICS transactions per day
80 percent are from the shop floor
40 transmissions per employee per day
1100 designs received per month
130 DNC facilities
600 CRT terminals
115 printers
55 bar code readers
10,000 control files distributed each month

12.3 DEERE & COMPANY

John Deere is recognized as a leader in the implementation of advanced manufacturing technologies.

In the mid-1970s, in response to decades of growth and the increased complexity of the manufacturing process, John Deere started a major modernization and reorganization program. Two new plants were constructed at the company's Waterloo, Iowa, facility. One plant was dedicated to manufacturing a line of diesel engines, the other for farm tractor assembly. The new plants freed up manufacturing responsibilities and space at the facility, and Deere implemented a reorganization of its other divisions based on group technology expertise. Hydraulics, drivetrains and special products were restructured as factories within factories

Deere experienced great success in reorganizing using the group technology philosophy. Part family cells replaced large functional process based departments. These new, smaller departments were dedicated to going from raw product to finished good within their unit, much like a mini-factory. Few materials changed hands from department to department, in marked contrast to the prior system.

The cellular organization led to improved efficiency and lowered operational costs. Excessive material handling was eliminated; prior to the adoption of the system, parts had followed a production path up to seven miles long. The manufacturing of complete parts within various cells eliminated long lead times. Quality improved as workers corrected production flaws while parts were still in the cell.

Deere's successes with group technology in the Waterloo plant led to the creation of the Deere Manufacturing Engineering Systems Department. In 1978, Deere began developing its own group technology system interviewing hundreds of its employees to determine what information they needed to identify part families based on similarities in material, geometric features and manufacturing process. Their recommendations led to the development of a CIM system that was implemented at a number of Deere plants worldwide.

The installation of CIM technology in a wide range of Deere operations earned the company recognition in the field. Deere received the first LEAD Award for Excellence in CIM from the Computer and Automated System Association of the Society of Manufacturing Engineers in 1981.

Later, the MES Department became the Computer Aided Manufacturing Services Division with the mission of establishing Deere as a leader in CIM.

The Department dedicated itself to solving the problems associated with the adoption of CIM on a wide scale. The group was charged with developing answers to the technological and structural changes required to implement CIM. Strategic analysis, long-term planning and corporate cultural changes were included in the group's responsibilities.

Based on the successes of the Computer Aided Manufacturing Services Division, Deere created Deere Tech Services in October of 1986. DTS offers the expertise in developing and implementing CIM to other companies looking for a strategic answer to manufacturing in an increasingly competitive environment. Through Deere Tech Services, the systems and problem-solving experience developed through the implementation of John Deere's successful CIM program are now commercially available to other companies looking for tangible, positive results. The division is currently working with clients such as Jaguar, Winnebago, Gates Rubber, Boeing, CAM-1, and the U.S. Air Force.

12.4 FROST, INC.

Frost, Inc. (Grand Rapids, MI), a manufacturer of conveyors and automation equipment, is an outstanding example of automation by a mid-size company. The company has been in business since 1915. With annual sales of $15- to $20 million, Frost initiated a three-year, $5.5 million automation project in 1984. A payback of 2.4 years is expected, and the target is to produce $225,000 in annual sales per employee. A major result is a 2,400 percent increase in product quality.

In order to implement the project, Frost formed an internal consulting firm, Amprotech, Inc., to create a business plan, design and install the automation equipment, and make it work. Amprotech is now marketing its skills in automation to small and medium size companies.

Chad Frost, chairman of Frost, Inc., directed the automation effort. His approach was not only to automate manufacturing, but to change the culture of the entire company. The automation plans were discussed with all employees, and they were given the opportunity to participate in some decisions affecting their work. For example, since the new plant was to run 24 hours per day, seven days per week, they were given the opportunity to establish a new work schedule. They came up with a plan to work three days on, four off in alternating 12-hour shifts.

The heart of the system is an AS/RS, a state-of-the-art piece of equipment designed and built by Transfer Technologies Inc., a subsidiary of Frost, Inc. It is an overhead, computerized electrified monorail. Powered carriers move hanging baskets, each with a load capacity of 500 pounds. As the powered carriers move in the track, each unit functions independently with its own motor on-board computer. There are ten power units capable of automatically changing speed on 700 feet of track. The number of units and amount of track are governed by the needs of the system. Each carrier will follow a closed-loop with spurs, switches and curves that allow it to pick up and deliver raw material and finished products throughout the plant.

The old plant, which is still operating side-by-side with the new facility, used ten forklift trucks; the new operation requires just one for both shipping and receiving functions. The small AS/RS system holds no more than two weeks' worth of raw inventory, and the size of the new plant requires just 29,000 square feet, compared with 75,000 square feet for the old plant. The side-by-side facilities have together cut the leadtime for trolley components from 16 to 20 weeks to two to four weeks. The quality improvement registered so far is 2,400 percent; the old plant had a rework rate of 24 percent; now the rate for both is 1 percent. Inventory occupies only 25 percent of its former space. From a peak of 185 employees in 1981, just over 100 remain as a result of attrition, automation, and a hiring freeze since the project began.

Already operational are five CNC machines, Mazak model QST-30, interfaced with five Roberts Corp. robots, all tied together through a Motorola 68000 microprocessor.

Software was written to integrate machines on a step-by-step microsecond basis, coordinating tools to process parts quickly. Individual tools are portable, meaning changes can be made on any machine with little effort. The computer systems in use are the IBM System 36, running on MAPICS software as the business host, and an IBM 4361 system running CADAM software for the engineering host.

12.5 GENERAL ELECTRIC

General Electric has a major effort to improve manufacturing productivity and maintain high quality through factory automation.

Not only has GE, companywide, been allocating $500 million annually to automate its own plants, but GE's industrial electronics business is also offering automation equipment and solutions to customers worldwide.

Not only has GE added about $600 million to $800 million in net income that the company would be in danger of losing if it hadn't moved to make its own plants more productive, but its showcase plants also serve as a demonstration of CIM technology for GE customers. This section examines one of these plants, the Appliance Park facility near Louisville, in depth, and provides an overview of two other GE CIM facilities.

The Louisville Plant

One of the best examples of an ultimate use of material handling equipment and systems is General Electric's dishwasher manufacturing plant at its huge Appliance Park complex in Louisville, KY.

A few years ago, GE was faced with a production operation that was highly manual. It included over 15 miles of an overhead, four-inch monorail system, with many manual transfers.

When GE began to redesign its line, one of the determinations was that the potential for damage due to the many manual transfer points had to be eliminated and a much higher raw and in-process inventory turnover had to be accomplished.

The major changes to the existing 780,000-square-foot plant had to be made without disrupting the manufacturing process. The major prerequisite was that, at no time, could the existing dishwasher plant be shut down.

Since GE wanted to achieve a fully integrated manufacturing facility to assemble its dishwashers, the alternatives consisted of two systems of delivery and assembly: synchronous and non-synchronous.

Work began by virtually tearing out major sections of plant No. 3 in the Appliance Park complex to make way for what was to become one of the most advanced automated assembly systems in the world. Essentially everything that one has ever heard of in terms of state-of-the-art manufacturing systems is embodied in the GE dishwasher operation.

Basically, what it includes is 3-1/2 miles of power and free conveyors, several automated transfer machines, nine robots, and a variety of specialized automated fabrication and assembly equipment, all under computer control.

The network of power and free conveyors is divided into four lines: one for tubs, a second for doors and two more for assembled dishwashers. Two of the lines merge in several areas. The third and fourth move units through final assembly lines, then to testing areas, and finally to packing areas, where the dishwashers are boxed and moved by conveyor to a warehouse.

The power and free system is controlled by three programmable controllers, which are linked to a minicomputer. The computer keeps inventory records and schedules work through the plant.

Each power and free carrier has an identifying bar code attached to it. Laser scanners throughout the plant read the codes and pass data to the computer, enabling it to track major in-process parts.

The factory has a total of 28 PCs, many of which have numerous data processing capabilities. Information is transferred between the computer and people via eight CRT terminals.

To help meet through-put requirements, GE uses automated transfer machines that work with the robots to transfer tubs to and from the power and free tub carriers. The transfer machines serve as interfaces between robots and carriers, enabling the robots to always be either loading or unloading tubs.

Doors are inspected and manually loaded onto power and free carriers, after being removed from molding machines. On demand, from one of the three PCs that control the conveyor, each carrier advances to the two-sided transfer station, where a robot takes the doors from their carriers and loads them on a special carrier chain conveyor.

At final assembly, an operator unloads each carrier and places the doors on door-assembly lines, where finishing touches are added manually. The completed doors then move on a monorail to a station about halfway down the final assembly line, where they are picked up manually and attached to tubs. Tubs, after being removed from molding machines, advance directly to overhead storage.

When called, each tub carrier moves to the firt station on the final assembly line. A robot unloads the tubs and a transfer machine places them into assembly carriers. The carriers progress from station to station, until doors are attached to the tubs and assembly is complete.

From the point where doors and tubs for the dishwashers are unloaded from molding machines and placed in carriers until the tubs and doors are transferred to final assembly, no human hand touches them.

What GE has accomplished by investing approximately $38 million in robotics, "point of use" subassembly manufacturing units, inventory control systems and other productivity-improving concepts, is an operation that now utilizes approximately the same number of employees as before, but produces a significantly higher number of dishwasher units than it did before the system was installed.

But, as importantly, these changes have improved manufacturing efficiency and quality so profoundly that sales of GE's dishwashers, which have won critical evaluations for quality and value, skyrocketed.

In specific terms, this state-of-the-art material handling operation has reduced the 28 separate transfers, or queuing operations, that existed on the old line down to three. This has played a key role in preventing in-process damage, thereby contributing to the quality of GE's dishwashers, simply because they are now handled less during production.

In regard to the just-in-time concept, the former process required five day's worth of tubs and doors to feed the line. Today, it requires less than one day's worth. That is an 80 percent reduction in raw and in-process inventory.

Also, GE attributes one-third of the improvement in increasing its dishwasher inventory turnover, from 12 times a year to over 24 times a year, to material handling.

Overall, GE's new dishwasher manufacturing operation has exceeded every one of its myriad goals. It is even perceived positively by GE's employees and, as a result of this joint cooperation, the company has been able to increase its dishwasher volume and market share.

Finally, the new dishwasher program is so successful, GE will continue its billion-dollar investment program to automate and upgrade the rest of its major appliance business.

Erie, PA

General Electric invested $316 million in upgrading productivity and quality at the Erie, PA, Transportation Systems Business Operations. Sixty-nine percent of the expenditure at the 5-million square foot, 7300-employee plant was for productivity and automation projects, and another 8 percent was for rearrangement of present facilities to enhance product flow, reduce cycles, and inventory control. The remainder was for additional floor space.

At the heart of the system is a flexible manufacturing system (FMS), installed by Giddings and Lewis, for machining a family of motor frames and gear boxes.

The system is capable of randomly machining parts and has the capacity to produce over 5,500 frames with an expansion capacity of 60 percent.

The FMS is comprised of: Nine heavy duty machine tools—two vertical milling machines, three horizontal machining centers, three heavy horizontal boring mills and one medium horizontal mill. All of these machines are equipped with GE1050CNC and automatic or robotic tool changers. Over 500 cutting tools are stored within the nine machines. The high usage cutting tools are automatically monitored and replaced by hookups as broken tool detection occurs. The machines are all tied together by a 212-foot long rail-tracked, chain-driven transporter which moves frames between 21 load/unload stations in the system.

Schenectady, NY

The CIM system has fully integrated marketing, engineering, manufacturing and finance functions at the small parts shop which annually produces some 325,000 parts in thousands of configurations. The parts are used in turbine-generators that are custom-manufactured to meet specific requirements of electric utility and industrial customers.

Technically, the CIM system's layered architecture is comprised of a Honeywell mainframe computer, GE's largest interactive graphics facility (CALMA), seven Data General MV8000 and MV10000 minicomputers, and hundreds of microcomputers. The key element in the network has been the creation of a unique part definition database called internally the Part Recognition Code (PRC). The code serves as a replacement for traditional drawings in terms of digitally defining the product for use by all business systems. The PRC is based on Group Technology Family of Parts concepts and provides all of the information, including notational data, normally provided by the engineering drawing.

A projected investment of $50 million by 1991, the CIM system will link all business functions in a common computer database. It uses a series of computers to process equipment orders from order entry through shipment without generating drawings and associated paperwork.

In the GE plant's small-parts shop, the prototype CIM installation, a computer database, contains about 25,000 unique part designs for the manufacture of small turbine components such as packing rings, spill strips, packing casings, oil deflectors, bolts, nuts, and studs. The shop turns out 350,000 parts per year to fill more than 15,000 orders from utility customers. It uses two dozen numerical control (NC) and 90 non-NC machine tools.

About 500,000 of 1.5 million original drawings are now stored on optical disks, inscribed and read using laser technology, and linked to remote drawing retrieval stations by microwave links.

12.6 GENERAL MOTORS

In the past few years, the General Motors Corp. has set out on an ambitious course of modernization. The company's goal is to transform its manufacturing operation from one based essentially on Henry Ford's assembly lines into an efficient, flexible process that will serve it well into the next century. Some of the steps that GM has taken in this regard, such as its acquisition of Electronic Data Systems and Hughes Aircraft, its joint effort with Toyota and its plans for a new Saturn factory, have been well publicized.

Increasingly, attention is also being paid to the key technological component of this modernization effort—GM's factory communications program, called Manufacturing Automation Protocol (MAP). General Motor's goal for MAP is to tie together all of the intelligent machines inside its factories. One indication of the scope of this endeavor is the fact that GM factories now house more than 40,000 intelligent devices, a number that is expected to increase between 400 and 500 percent over the next five years.

Bringing communications to the factory floor is an absolute prerequisite for all of GM's modernization plans. Consider, for example, GM's intentions for its Saturn line of cars. In a few years, Saturn customers will be able to walk into a car dealership and, by working their way through a list of features on a computer terminal, choose every detail of the car they want to drive. Once that happens, the order will be relayed to the Saturn factory, where the car will be custom-made, and then shipped back to the customer within a week.

The Saturn plant will be able to make as many as 500,000 cars a year in this manner—twice as many as conventional plants, which have none of the flexibility that the Saturn facility will have. For this to happen, dealers, inventories, order and entry systems, the design department, the factory floor, the shipping department and other parts of the company must be tied together, which would be impossible without the extensive inter-organizational communications that MAP will bring about.

The numbers involved in this effort are staggering. GM is the world's largest automation customer, and expects to spend about $40 billion on automation before 1990. In fact, MAP was the result of GM's conclusion that it simply could not afford to let the field remain unstandardized—that it would have to act, unilaterally if necessary, to bring some order to automation communications. The alternative for the company would have been to face the staggering costs involved in trying to get incompatible machines from its hundreds of different vendors to work together.

GM's actions have been welcomed by virtually all industry leaders in manufacturing and electronics, since they, too, have nothing to gain and much to lose from an unstandardized market. The MAP program has been endorsed by IBM, Intel, DEC, Ford, Allen-Bradley, Gould, Boeing, Westinghouse, GE, Kodak, Dupont, Volkswagen, Renault and many others. All of them belong to a MAP users group which is working on details of the program. A special panel of the IEEE—the 802.4 subcommittee—is developing MAP standards, as is the National Bureau of Standards. There is also support for MAP from Canada and Europe. Clearly, MAP is on the verge of becoming a global standard for industrial automation, and has given a major impetus to the field, bringing orderly growth to a market in a way that only standards can.

12.7 MERCK & COMPANY, INC.

Merck is a multi-national pharmaceutical company, with annual sales on the order of $3.5 million. Merck's bulk manufacturing facilities produce the chemicals that are the active ingredients of their drugs, typically in batch mode, either by chemical synthesis or by fermentation or, sometimes, by a combination of both.

About twenty years ago, Merck became a pioneer in the pharmaceutical industry when it undertook to develop computer control for certain of its batch processes, particularly for multi-batch, multi-process bulk chemical facilities which were then being contemplated and were built in the early- and mid-1970s. Eventually, they also applied computer control to a variety of pharmaceutical manufacturing operations, both here and abroad.

In the mid-1970s, Merck's Automation and Control Group developed their own process control system and user-oriented language, called "AUTRAN" (it is still in use). The firm installed a number of fairly large, central control systems first in chemical plants, then in pharmaceutical plants in the United States, Ireland, England, and more recently, in Japan.

However, they were not completely alone. Their knowledge of what colleagues and competitors do is, of course, imperfect, but they did know that in the early-1970s many firms were beginning to use computers in the laboratory, and eventually learned that one was very active in applying computer assistance to fermentation manufacturing processes, an aspect of Merck's operations that was addressed only recently.

Process control systems, both those developed internally and, eventually, those purchased from various vendors, are in place in many chemical and pharmaceutical plants in the United States, Japan, and Europe. There are numerous uses of programmable controllers for the sequencing of local operations. The Automation & Control Group is now a corporate area with a staff of about 80 people. The company's divisions are in the process of strengthening their capability to install and support automation. Process control is now only one part of the Automation & Control Group's effort. About half of the activity is in support of laboratory automation in Merck's research and quality control labs. Merck branched out several years ago into automated warehousing with the installation of three high-bay warehouses in Europe, and is now moving into other kinds of computer assistance in existing conventional warehouses in the United States. The high-bay warehouses are highly automated in the physical sense. The current approach revolves around the exchange and recording of information and directives using radio-transmission computer terminals, without the need for a great deal of mechanization. The company is now beginning its first project with that technology in their pharmaceutical plant at West Point, PA. They are also running pilot efforts with industrial robotics on a packaging line. The prospects for machine-vision systems show great promise, given the heavy inspection burden on these lines. In May 1985, they are installing the first two machine-vision systems on packaging lines.

The use of laboratory robots is showing more vigor at Merck than use of the industrial variety. They have about 15 in analytical and quality control labs, and the number seems to be rising rapidly. Merck has also made an initial probe into the world of artificial intelligence, with the purchase of specialized hardware and tools for developing a simulation package to help their chemical engineers optimize the designs of new processes.

12.8 TEXAS INSTRUMENTS

The Texas Instruments' Defense Systems & Electronics Group plant at Sherman, TX, is TI's most highly automated and vertically integrated government electronics factory. From design to fabrication and assembly, integrated computer systems are a critical part of the process.

During the past six years the TI Sherman plant has realized a 60 percent increase in productivity, a 60 percent increase in on-time deliveries from vendors, and a 47 percent decrease in work in process (WIP). Cycle times have been reduced by 45 percent and major cost savings have been realized by avoiding increased factory and warehouse capacity.

The need for integrated information systems begins when the design engineer conceives a new product. The engineering information systems assist in the development process.

When TI receives a new order for equipment a project structure is set up. This can be transferred directly from the engineering information systems to the manufacturing systems, or it can be input manually.

Information is extracted from these inputs and from the inventory systems by the material requirements planning system to generate the time phased material plan.

Upon approval of the appropriate manufacturing engineer, electronic requests for purchase and fab purchase orders are released to buyers and fab shops. In cases where option agreements have been established, purchase orders may go directly to a supplier's order entry system electronically.

Meanwhile, computer-aided design systems with three-dimensional surfaced modeling capabilities are used to design parts and fixturing. The models are also used to prove toolpaths before "cutting chips."

The computer-aided design data is converted to N-C programs and transferred to a site computer. The programs are requested through the terminal for remote interface control systems, or TRICS by the machine or cell operator.

The TRICS unit stores the latest revision of the programs which will be used by the numerical control machines connected to it. This ensures that the machining operations can continue even if communications to the host computer are lost. An internally mounted hard disk stores programs in the TRICS so programs are not lost if the TRICS is powered off. This integration from design through machining allows us to produce useable parts the first part off the machine.

While the CAD-to-machining link is a highly visible portion of the factory integration achieved at TI Sherman, most of the information sharing and transfer is done in the major systems which control labor, accounting, inventory, automated warehousing, purchasing, receiving, material planning, scheduling, quality control, and work order routing generation.

Once the manufacturing schedule is approved the work orders are released by production control. This automatically opens accounts for labor and material.

As material is received, it is assigned to barcoded carriers for automated movement and tracking. It is then checked for size and weight to ensure it is within the material handling system's capability. As parts progress through the manufacturing cycle the material plan is constantly updated so corrective action can be taken to prevent waste or disruptions in the factory.

Warehouse and floor control systems automatically store and retrieve parts from an eight-story warehouse with over 6,000 storage locations.

The process floor control system (or P-F-C-S) was developed by TI to provide the closed-loop scheduling necessary for "hybrid just-in-time" manufacturing. P-F-C-S ensures the accurate scheduling of material to the correct operation based on need.

The system interprets the schedule, parts availability, and demand at downstream operations to automatically dispatch material.

Barcode entry is used by the operator to update several systems, including the status on P-F-C-S. This keys the floor control systems to pick up the finished parts, and to dispatch a new pallet of material if needed. Barcoding also controls inputs to the labor system which, in turn, feeds performance reporting the cost control systems.

Process floor control system tracks material against the schedule all the way to final assembly. It creates demand at upstream operations as if it were an electronic KAN BAN system.

At TI Sherman, operators are responsible for inspecting 100 percent of their own work. A large investment in training, equipment, and tools has been made to give operators the means to identify trends and to take action to prevent defects. Quality control inspectors monitor the output and audit the process.

Texas Instruments at Sherman believes that the operator is the expert on how the job gets done, and is the best person to identify and prevent problems. They also believe, however, that most of the improvements that can be made in the manufacturing process are beyond the operator's control.

In keeping with that knowledge, systems are being developed to feed process information from the shop floor. Connecting these systems to the network will allow engineering and management to better understand the process capabilities of the shop so that better design and manufacturing decisions can be made.

The lessons and techniques learned at the Texas Instruments plant in Sherman have already been used in the automation of other sites.

12.9 3M

3M's 45 major product lines share common origins in a diverse technology base. Important 3M technologies such as precision coating and bonding, biochemistry, electronics and imaging are being combined with and supplemented by new technologies with excellent potential for generating proprietary new products.

In today's competitive business environment, it is not enough for a company just to have a strong flow of innovative products. These products must also be produced efficiently.

New concepts in manufacturing are helping 3M stay competitive and maintain above-average profitability through reduced costs. Among these concepts are alternatives to costly petroleum-based solvents in the manufacturing process.

Today, innovative techniques which eliminate use of solvents in coatings are being used to produce a variety of pressure-sensitive tapes. These solvent-free processes not only are more economical to use, but are resulting in greater environmental safety and higher performance characteristics.

3M people are also working to reduce costs by applying leading-edge manufacturing technologies and systems. Examples include:

1. In the fast-growing video-cassette market, automated equipment helps control the manufacturing process from start to finish. Numerous complex operations are performed in seconds at a single work station, and electronic intelligence monitors every stage of the manufacturing process to insure the highest possible quality.

2. In the production of polyester base film, which is used in 3M products ranging from computer tape to photographic film, a new automated production line is among the fastest and most efficient in the world. Process settings are microprocessor controlled, and deviations during film manufacture are compensated for automatically.

3. Computer-controlled equipment automates the production of parts used in surgical cutting and drilling tools. Previously, a dozen or more separate steps were required to produce finished parts. Now, thanks in large measure to automatic tool-changing capabilities, the process has been greatly streamlined, with just a few steps required.

More consistent quality and reduced inventory levels are among the ways newer manufacturing technologies are helping increase efficiency. Combined with the efforts of employees and raw material suppliers, automated equipment helps reduce waste, rework and other errors, which can substantially increase product costs. The cost of carrying inventory also is being reduced because, with newer technologies, products can be produced in smaller quantities as they are needed, rather than in extended production runs. This saves both on floor space and working capital.

3M's distribution organization has a simple goal: deliver products to the end-user at the lowest cost. But the large scale of 3M's operation makes cost reduction no simple matter. From over 100 manufacturing plants nationwide, a huge variety of products funnel into a single distribution center and, from there, to nine regional warehouses. Recently, however, 3M found an effective way to cut costs. By implementing bar codes throughout the distribution channel, the company estimates total annual savings of $16 million from productivity and efficiency.

The first step meant adopting a bar code that could be used by the manufacturers, distributors, wholesalers and retailers. A single code prevents expensive cross-reference files and additional labeling costs. 3M chose the Uniform Product Code (UPC) as its standard. Suppliers labeled all product containers with the 14-digit code identifying the manufacturer, country of origin, and the item.

The UPC code allows electronic data communication to handle almost every aspect of order and inventory control. For example, a computer can automatically replenish an inventory item. It creates a purchase order and transmits the order directly to the manufacturer's system. The elimination of manual transcription, key entry, and telephone time means greater productivity. In addition, the common errors of manual input are eliminated. For 3M, enhanced productivity and error reduction offered a savings potential of $6.2 million a year.

3M also discovered that internal warehouse replenishment orders achieved a new level of speed and accuracy—direct results of the bar code system. Due to better product identification, inventory turnover increased dramatically. The bottom line was a $5 million reduction in inventory carrying costs. And the incidence of late orders or other problems connected with out-of-date or erroneous inventory records dropped sharply. 3M estimates it gained $1.4 million in customer service productivity. When fully implemented by 3M in all its warehouses, the company expects to save $3.4 million in faster throughput and order turnover.

A real-time, integrated MRP II system has reduced inventory 66 percent and more than doubled throughput for the 3M Company's prototype shop. Jobs that once took 10 to 14 days to complete now only take 4 to 5. ASK Computer Systems software was used.

In the past, as many as six different automated systems had been used in the St. Paul, MN shop, which performs machining, metal fabrication, and assembly functions on 800 to 1,000 jobs each month. None of the systems were integrated, and often jobs were not turning around quickly enough to satisfy the demands of the engineering department.

The system keeps stock moving while improving parts tracking. Better inventory control has resulted in a jump in on-time delivery rates from 55 to 95 percent. Backlogs have been virtually eliminated, too. The shop once had a 2 to 4 day backlog in processing parts shipments, and now it's 2 to 4 hours.

An in-house application of machine vision at 3M is for automated analysis of polymer mixing. Every polymer processing line includes some stage of mixing. The best technique for measuring mixing quality in the early stages of the mixing process is to determine the statistical distribution of the striation length and width histograms. Direct methods of performing this measurement are labor-intensive. An image analysis technique that mimics the direct measurement method and computes the length/width histograms was developed at 3M by R.D. Iverson.

12.10 WHIRLPOOL

Whirlpool operates in an industry that consistently ranks in the top 10 of the 60 industries whose productivity is monitored by the U.S. Department of Labor. Sales per employee at Whirlpool are approximately $140,000, annually, and rising.

Whirlpool's four point manufacturing strategy is:

1. Integrating manufacturing
2. Automating for efficiency
3. Designing for manufacturability
4. Improving plant utilization

Integrating Manufacturing

Whirlpool is strengthening its long-term competitiveness through streamlined plants, leading edge technology and products designed to benefit from the latest manufacturing techniques. Further enhancing the company's competitive position is a new, computer-based, state-of-the-art information system—Whirlpool Manufacturing Control System (WMCS)—that can quickly gather and integrate information about all of the systems that support manufacturing into a comprehensive, easy-to-understand overview.

Whirlpool Manufacturing Control System was phased-in over three years and fully operational in 1987, replacing three, group-based information systems that functioned well but independently.

Because it can provide a real-time picture of what is happening on Whirlpool factory floors, WMCS allows better management of materials and labor. The system permits coordinating bills of material, centralizing purchasing activity and responding to customer product needs more quickly. Cost savings result from reduced inventory levels, reduced scrap levels, reduced parts obsolescence, better labor utilization and more cost-effective purchasing.

Because raw material and component purchases from outside suppliers consume more than half of the sales dollar, the long-term positive impact of WMCS in the purchasing area already is very significant. WMCS permits coordinated buying by consolidating data on purchasing needs throughout the company, item by item. Along with the economies of scale gained from increased volume purchasing, operational economies follow from dealing with fewer part numbers and fewer suppliers. In addition, having sufficient, timely information about procurement requirements enables the company to extend its sourcing to a growing number of highly competitive international suppliers.

The cost of developing and implementing WMCS will be approximately $15 million. In return, that investment will generate annual savings estimated at upwards of $30 million.

Automating for Efficiency

Along with having streamlined, highly efficient plants, Whirlpool must strive for continuing improvements in manufacturing productivity if it is to remain competitive in tomorrow's marketplace. This is being done through increased automation and continual equipment upgrading on the factory floor.

Examples abound in Whirlpool plants of cost and time savings achieved by upgrading from manual to automated and increasingly sophisticated manufacturing, assembly and inspection techniques. Automation also improves product quality and assemblies with exacting consistency.

A typical example of benefits achieved through automation is the range cabinet line at the Findlay Division. Converting from manual to automated production has reduced labor time by about 70 percent, increased units per hour by approximately 35 percent, reduced scrap costs per unit by over 75 percent and doubled overall productivity as measured by cabinets produced per employee per hour.

Another illustration also can be found at the Findlay Division. A robot performs the "seemingly simple" task of transferring a 40-pound dishwasher tub from one conveyor to another. In the process, however, it must search a furnace conveyor with an optical sensor until it locates the desired tub. It then must synchronize itself with the moving line, lift the tub from its hanger and transfer it to an assembly conveyor. The entire operation is computer-controlled, and is done in less than 10 seconds. Employees no longer are required to manually lift hot (300 °F), heavy tubs from one line to another.

While automation often reduces labor requirements, it also provides new and more challenging opportunities for which Whirlpool people are being trained. Areas include designing, programming, and maintaining new and upgraded equipment.

Designing for Manufacturability

Designing and redesigning products to gain maximum economic advantage from the latest manufacturing techniques is yet another route being followed by Whirlpool to improve productivity and cost competitiveness.

The company's best and most recent example of the advantages of designing/redesigning for optimum "manufacturability" is found at the Clyde, Ohio Division. A major product design and manufacturing revolution is focusing on a product—the automatic clothes washer—that, until now, has been unchanged in any fundamental way for 30 years.

What motivated Whirlpool to undertake the phase-out of the 30-year old washer designs that are very reliably satisfying customers to this day? Not surprisingly, rising energy and raw materials prices plus the cost of updating older plants with new technology were seen as increasing production costs. The need became apparent for new product designs of equal or better quality that could be manufactured more economically.

The "Design 2000" automatic washer line being produced at Clyde clearly meets that need. Representing a totally new manufacturing approach for the company, it is the result of years of research, development and testing by engineers, designers and consumer affairs specialists. Under the new system, for example, the previous "outside-in" method of washer assembly—using the cabinet as the structural support frame—is replaced with an "inside-out" approach that starts from the base and works up. This allows easy and complete testing for quality before the cabinet encloses the machine.

In all, manufacturing of the "Design 2000" uses 15 new and unique production processes. Benefits resulting from these manufacturing design innovations include improved product performance and reliability, easier serviceability and a savings in both labor and manufacturing energy costs.

Improving Plant Utilization

Assuring that Whirlpool remains on the leading competitive edge in tomorrow's marketplace is a corporate-wide mission of the highest priority. This mission is being addressed, in part, by a major plant consolidation and reorganization program now well underway to increase productivity and cost competitiveness.

Launched in 1983, the program was prompted by a facilities assessment study that confirmed that certain plants were operating at high efficiency but that others were underutilized, too old to be competitive, or otherwise were not well equipped for top-quality, cost-effective, high-volume production. The study also confirmed that some products being manufactured in more than one plant would be produced more efficiently in a single plant.

These findings triggered several important management decisions in the past two years, including:

> Consolidation of all air control products manufacturing at the company's wholly owned subsidiary, Heil-Quaker Corporation
>
> Phasing out automatic washer production at the St. Joseph, MI plant and dedicating that plant exclusively to parts manufacture
>
> Consolidation of all automatic washer production at the Clyde, OH Division
>
> Moving vacuum cleaner production from the St. Paul, MN Division to the Danville, KY Division
>
> Moving freezer and residential ice-maker production from St. Paul to Evansville—then closing the St. Paul Division
>
> Consolidating dryer production at the Marion, OH Division and range production at the Findlay, OH Division

Some of these decisions have already been implemented. As might be expected, the cost of these ambitious efforts has been substantial, penalizing earnings in the short term. The resulting long-term benefits, however, are expected to more than justify the costs incurred—chiefly through a 15 percent reduction in manufacturing space plus significant reductions in employment and other direct costs for comparable production levels.

Appendix **1**

Glossary

Accuracy: Quality, state, or degree of conformance to a recognized standard or specification.

Algorithm: A procedure for solving a problem in a finite number of steps.

Analog: An expression of values which can vary continuously, e.g., translation, rotation, voltage, or resistance.

AND/OR Graph: A generalized representation for problem reduction situations and two person games. A tree-like structure with two types of nodes. Those for which several successors of a node have to be accomplished (or considered) are AND nodes. (In about half the literature the labeling of AND and OR nodes is reversed from this definition.)

AND, OR, AND STREAM PARALLELISM: Different techniques for implementing parallel operations, based on "and, or" and pipelines execution hierarchies.

Argument Form: A reasoning procedure in logic.

Artificial Intelligence (AI): (1) A discipline devoted to developing and applying computational approaches to intelligent behavior. Also referred to machine intelligence or heuristic programming; (2) The study of human intelligence through the use of methods and metaphors depending essentially on computers and computation; (3) The computational study of tasks and processes which have previously been observed only in association with human intelligence.

Artificial Intelligence (AI) Approach: An approach that has its emphasis on symbolic processes for representing and manipulating knowledge in a problem solving mode.

ASCII: Abbreviations for American Standard Code for Information Interchange. It is an eight-bit (7 bits plus a parity bit) code for representing alphanumerics, punctuation marks, and certain special characters for control purposes.

Automatic Programming: System in which a program specification or design is automatically translated into a working program with minimal operator intervention.

Automation: Automatically controlled operation of an apparatus, process or system by mechanical or electronic devices that take the place of human observation, effort and decision.

Autonomous: A system capable of independent action.

Backtracking: Returning (usually due to depth-first search failure) to an earlier point in a search space in an expert system. Also a name given to depth-first backward reasoning.

Backward Chaining: A form of reasoning in an expert system starting with a goal and recursively chaining backwards to its antecedent goals or states by applying applicable operators until an appropriate earlier state is reached or the system backtracks. This is a form of depth-first search. When the application of operators changes a single goal or state into multiple goals or states, the approach is referred to as problem reduction.

Bandwidth: The number of cycles per second expressing the difference between the lower and upper limiting frequencies of a frequency band.

Bottom-Up Control Structure: A problem-solving approach that employs forward reasoning from current or initial conditions. Also referred to as an event-driven or data-driven control structure.

CAD: An acronym for Computer-Aided Design.

Calibration: Reconciliation to standard of measurement.

CAM: An acronym for Computer-Aided Manufacturing.

CCTV: Abbreviation for Closed Circuit Television. A television system that does not broadcast TV signals but transmits them over a closed circuit.

Certainty Factor (*CF*): A numeric value that indicates a measure of confidence in the value of a parameter on the part of the consultant or the client.

Charge-Coupled Device (*CCD*): A self-scanning semiconductor array that utilizes MOS technology, surface storage and information transfer by digital shift register techniques.

Charge-Injection Device (*CID*): A conductor-insulator-semiconductor structure that employs intracell charge transfer and charge injection to achieve an image sensing function.

Code Reading: Actual recognition of alphanumerics or other set of symbols, e.g., bar codes, UPC codes.

Code Verification: Validation of alphanumeric data to assure conformance to qualitative standard-subjective.

Cognition: An intellectual process by which knowledge is gained about perceptions or ideas.

Cognitive Science: A general term covering all those branches of science concerned with the study of thinking, including artificial intelligence, psychology, and linguistics.

Combinatorial Explosion: The rapid growth of possibilities as the search space expands. If each branch point (decision point) has an average of n branches, the search space tends to expand as n-d, as the depth of search, d, increases.

Common Sense: The ability to act appropriately in everyday situations based on one's lifetime accumulation of experiential knowledge.

Common Sense Reasoning: Low-level reasoning based on a wealth of experience.

Compatibility: The compatibility of using an instruction, program, or component on more than one computer with the same result.

Compile: The act of translating a computer program written in a high-level language (such as LISP) into the machine language which controls the basic operations of the computer.

Computational Logic: A science designed to make use of computers in logic calculus.

Computer-Aided Design (CAD): The use of a computer to assist in the creation or modification of a design.

Computer-Aided Manufacturing (CAM): The use of computer technology in the management, control, and operation of manufacturing.

Computer Architecture: The manner in which various computational elements are interconnected to achieve a computational function.

Computer Graphics: Visual representations generated by a computer (usually observed on a monitoring screen).

Computer Network: An interconnected set of communicating computers.

Control Structure: Reasoning strategy. The strategy for manipulating the domain knowledge to arrive at a problem solution.

Convolve: The process of superimposing an m × n operator over an m × n pixel area (window) in the image, multiplying the corresponding points together and summing the result.

Correlation: Process of assessing the relationship between two or more models of an image.

Data: A general term for any type of information.

Database: A large collection of records stored on a computer system from which specialized data may be extracted, organized, manipulated by a program. Any organized and structured collection of data in memory.

Data Base Management System: A computer system for the storage and retrieval of information about some domain.

Data-Driven: A forward reasoning, bottom-up problem solving approach.

Data Structure: The form in which data is stored in a computer.

Debugging: Correcting errors in a plan.

Decision Tree: A way of representing a complex series of choices as a branching diagram looking like a tree.

Decision Support Systems: Computer systems intended to provide managers and executives conveniently with information useful to making business decisions, e.g., sales figures, market projections, a means of carrying out "what-if" calculations, scheduling, and expert advice. Usually based on mathematical techniques.

Deduction: A process of reasoning in which the conclusion follows from the premises given.

Digitizing (digitization): Process of converting an analog video image into digital brightness values that are assigned to each pixel in the digitized image.

Disproving: An attempt to prove the impossibility of a hypothesized conclusion (theorem) or goal.

Domain: The problem area whose solution is addressed by the knowledge base and the inference engine; the area of interest, e.g., bacterial infections, prospecting, VLSI design.

DOS: Disk Operating System. The accepted standard operating system for the IBM PC. Microsoft (MS) developed the system under IBM's direction, hence the use of MS-DOS. IBM now markets its own PC-DOS. Most applications for the PC, including AI programs, run under DOS.

Editor: A software tool to aid in modifying a software program.

Embed: To write a computer language on top of (embedded in) another computer language (such as LISP).

Emulate: To perform like another system.

Engineering Workstation: A type of computer with a high-quality display originally intended for engineering design but now much used for expert systems.

Equivalent: Has the same truth value (in logic).

Evaluation Function: A function (usually heuristic) used to evaluate the merit of the various paths emanating from a node in a search tree.

Expert System: A computer program that uses knowledge and reasoning techniques to solve problems normally requiring the abilities of human experts. Because their functioning relies heavily on large bodies of knowledge, expert systems are sometimes known as Knowledge-Based Systems. Since they are often used to assist the human expert, they are also known as Intelligent Assistants.

Expertise: The set of capabilities that underlies the high performance of human experts, including extensive domain knowledge, heuristic rules that simplify and improve approaches to problem-solving, metaknowledge and metacognition, and compiled forms of behavior that afford great economy in skilled performance.

Explanation Facility: A feature of many expert systems that tells what steps were involved in the process by which the system arrived at a solution. These facilities can be simple traces of steps, or they can be more complex, supplying encoded reasons why the solution uses one alternative rather than another.

Exploratory Programming: A set of techniques developed by the artificial intelligence community to deal with problems for which the design of a solution cannot be known in advance. A major premise of this type of programming is that, by exploring various techniques and attempting to rapidly prototype a solution to some small subset of the problem, it can be better understood. Supporting techniques include interactive editing and debugging, integrated programming environments, and graphics-oriented user interfaces.

Factory: A manufacturing unit consisting of two or more centers and the materials transport, storage buffers, and communications which interconnect them.

Fail Safe: Failure of a device without danger to personnel or damage to product or plant facilities.

Fault Diagnosis: Determining the trouble source in an electro-mechanical system.

Features: Specific data elements describing the image content of a scene, such as edge point locations, centroid, etc.

Fiber Optics: A communication technique where information is transmitted in the form of light over a transparent material (fiber) such as a strand of glass. Advantages are noise free communication not susceptible to electromagnetic interference.

Fifth Generation: The era of technology (from today's standpoint) that heralds the next major technical advancements. Included in the fifth generation technologies are new super computers, commercial AI applications, improved storage devices, and larger scale integrated circuits.

Flexible Manufacturing: Production with machines capable of making a different product without retooling or any similar changeover. Flexible manufacturing is usually carried out with numerically controlled machine tools, robots, and conveyors under the control of a central computer.

Forward Chaining: In an expert system, a method of determining a parameter's value by evaluating the action of a rule when the premise of the rule is true. This method does not cause tracing of the parameters within the premise of the rule. Event-driven or data-driven reasoning.

Frame: A data structure for representing stereotyped objects or situations. A frame has slots to be filled for objects and relations appropriate to the situation.

Front-End: An information processor serving as an interface between a communications terminal and a data processing system.

Functional Application: The generic task or function performed in an application.

Fuzzy Logic: A way of reasoning approximately using indicators such as "true," "not very true," "many" and "few."

Fuzzy Set: A generalization of set theory that allows for various degrees of set membership, rather than all or none.

Goal: A condition or set of conditions to which a valid solution must conform.

Goal Driven: A problem-solving approach that works backward from the goal.

Goal Regression: A technique for constructing a plan by solving one conjunction subgoal at a time. Checking to see that each solution does not interfere with the other subgoals that have already been achieved. If interferences occur, the offending subgoal is moved to an earlier non-interfering point in the sequence of subgoal accomplishments.

Graph: A set of nodes connected by arcs.

Heuristics: A technique or assumption that is not formal knowledge, but which aids in finding the solution to a problem; rules of thumb or empirical knowledge used to help guide a problem solution.

Heuristic Search Techniques: Graph searching methods that use heuristic knowledge about the domain to help focus the search. They operate by generating and testing intermediate states along potential solution paths.

Hierarchical Approach: An approach to vision based on a series of ordered processing levels in which the degree of abstraction increases as we proceed from the image level to the interpretation level.

Hierarchical Planning: A planning approach in which first a high-level plan is formulated considering only the important (or major aspects. Then the major steps of the plan are refined into more detailed subplans.

Hierarchy: A system of things ranked one above the other.

Higher Order Language (HOL): A computer language (such as FORTRAN or LISP) requiring fewer statements than machine language and usually substantially easier to use and read.

Human Engineering: The task of designing human-machine interfaces to achieve effective human utilization of machine capacities.

Human Interface: A subsystem of any computing system with which the human user deals routinely. It aims to be as "natural" as possible, employing language as close as possible to ordinary language (or the stylized language of a given field) and understanding and displaying images, all at speeds that are comfortable and natural for humans. The other two subsystems in an expert system are the knowledge-based management subsystem and the inference subsystem.

Icon: A symbol to which a computer user can point an interface device in order to select a function, such as "move window."

Identification: Determination of the identity of an object by reading symbols on the object.

Image Analysis: Process of generating a set of descriptors or features on which a decision about objects in an image is based.

Image Interpretation: Process of building a description of a scene and then matching it against symbolic prototypes stored in a computer memory.

Image Preprocessing: Reduction of image data to a manageable form.

Image Processing: The examination by a computer of digitized data about a scene and the features in it in order to extract information.

Image Understanding (IU): Visual perception by a computer employing geometric modeling and the AI techniques of knowledge representation and cognitive processing to develop scene interpretations from image data. IU has dealt extensively with three-dimensional objects.

Implies: A connective in logic that indicates that if the first statement is true, the statement following is also true.

Individual: A non-variable element (or atom) in logic that cannot be broken down further.

Inference: The process of reaching a conclusion based on an initial set of propositions, the truths of which are known or assumed.

Inference Engine: In an expert system, the part of the software structure that applies the knowledge base expertise to the client's information to infer a solution to the problem. Another name given to the control structure of an AI problem solver in which the control is separate from the knowledge.

Integration: (1) A software design concept that allows users to move easily between application programs, or to incorporate data from one program into another, such as moving data displayed in a graphics program into a text documentation; (2) A computing approach that allows an organization to match its communications and information needs across organizational levels to specific products (systems) through the use of common system and information architectures.

Intelligence: The degree to which an individual can successfully respond to new situations or problems. It is based on the individual's knowledge level and the ability to appropriately manipulate and reformulate that knowledge (and incoming data) as required by the situation or problem.

Intelligent Assistant: An AI computer program (usually an expert system) that aids a person in the performance of a task.

Intelligent Front-End: A human-computer interface that uses knowledge-based techniques to provide a more powerful and helpful service to the user.

Intelligent Robotics: More flexible and powerful robotics systems than conventional ones that simply follow a set of preprogrammed movements.

Intelligent System: A system equipped with a knowledge base that can be manipulated in order to make inferences. The distinction between a system that can perform intelligent operations and one that merely manipulates data can be characterized in a highly simplified example.

Interactive Environment: A computational system in which the user interacts (dialogues) with the system (in real time) during the process of developing or running a computer program.

Interface: The system by which the user interacts with the computer. In general, the junction between two components.

Job Shop: A discrete parts manufacturing facility characterized by a mix of products of relatively low volume production in batch lots.

Knowledge: Facts, beliefs, and heuristic rules.

Knowledge Acquisition: The extraction and formulation of knowledge derived from extant sources, especially from experts.

Knowledge Base: AI databases that are not merely files of uniform content, but are collections of facts, inferences and procedures, corresponding to the types of information needed for problem solution.

Knowledge Base Management: Management of a knowledge base in terms of sorting, accessing and reasoning with the knowledge.

Knowledge Base Management System: One of three subsystems in an expert system. This subsystem "manages" the knowledge base by automatically organizing, controlling, propagating, and updating stored knowledge. It initiates searches for knowledge relevant to the line of reasoning upon which the inference subsystem is working. The inference subsystem is one of the other two subsystems in an expert system; the third is the human interface subsystem with which the end-user communicates.

Knowledge-Based System/Expert System: A program that allows computers to draw conclusions from a knowledge base (as opposed to a data base) that has been structured from experts' rules in a given field (e.g., medicine, geology, financial planning), giving the user the benefit of many experts' knowledge and experience in machine form.

Knowledge Engineer: The person who specializes in designing and building expert systems in conjunction with the consultant.

Knowledge Engineering: The AI approach focusing on the use of knowledge (e.g., as in expert sytems) to solve problems.

Knowledge Representation (KR): The form of data-structure used to organize the knowledge required for a problem.

Knowledge Source: An expert system component that deals with a specific area or activity.

LISP: List Programming. A computer programming language preferred by AI researchers and developers for working with AI techniques.

LISP Interpreter: A part of many LISP-based software tools that allows specific list-processing operations such as match, join, and substitute, to execute on a general-purpose computer rather than a special-purpose LISP machine.

LISP Machine: A single-user workstation with an architecture dedicated to the efficient writing and execution of applications using the LISP programming language.

List: A sequence of zero or more elements enclosed in a pair of parentheses, where each element is either an atom (an indivisible element) or a list.

Logical Operation: Execution of a single computer instruction.

Logical Representation: Knowledge representation by a collection of logical formulas (usually in First Order Predicate Logic) that provide a partial description of the world.

Logic Programming: The technique of declarative programming using the principles of classical logic. Prolog is probably the best known implementation.

Machine Learning: Techniques whereby computers can learn from their own experience rather than being told. A chess-playing machine, for instance, can learn a particular move leads to defeat and so remember not to make it again.

Machine Vision: Processing of producing useful symbolic descriptions of a visual environment from image data—automatic interpretation of imagery to control a manufacturing process.

Microcode: A computer program at the basic machine level.

Natural Language Interface (*NLI*): A system for communicating with a computer by using a natural language.

Natural Language Processing (*NLP*): Processing of natural language (e.g., English) by a computer to facilitate communication with the computer, or for other purposes such as language translation.

Natural Language Understanding (*NLU*): Response by a computer based on the meaning of a natural language input.

Node: A point (representing such aspects as the system state or an object) in a graph connected to other points in the graph by arcs (usually representing relations).

Object-Oriented Programming: A programming approach focused on objects which communicate by message passing. An object is considered to be a package of information and descriptions of procedures that can manipulate that information.

Operators: Procedures or generalized actions that can be used for changing situations.

Overload: A load greater than that which a device is designated to handle.

Parallel Processing: Simultaneous processing, as opposed to the sequential processing in a conventional (Von Neumann) type of computer architecture; processing more than one program at a time through more than one active processor.

Parameters: The facts that take on specific values during a consultation (variables). Parameters are the predefined attributes of the problem domain.

Pattern Matching: Matching patterns in a statement or image against patterns in a global data base, templates, or models.

Pattern Recognition: The process of classifying data into predetermined categories.

Perception: An active process in which hypotheses are formed about the nature of the environment, or sensory information is sought to confirm or refute hypotheses.

Personal AI Computer: New, small interactive, stand-alone computers for use by an AI researcher in developing AI programs. Usually specifically designed to run an AI language such as LISP.

Plan: A sequence of actions to transform an initial situation into a situation satisfying the goal conditions.

Portability: The ease with which a computer program developed in one programming environment can be transferred to another.

Predicate Logic: A modification of Propositional Logic to allow the use of variables and functions of variables.

Premise: A first proposition on which subsequent reasoning rests.

Probability: The mathematical way of describing likelihood, as a ratio similar to "odds."

Problem Reduction: A problem-solving approach in which operators are used to change a single problem into several subproblems (which are usually easier to solve).

Problem-Solving: A procedure using a control strategy to apply operators to a situation to try to achieve a goal.

Procedural Programming: The conventional way of driving a computer by giving it a sequence of instructions, as opposed to declarative programming.

Programmable Controller (PC): A solid-state control system which has a user programmable memory for storage of instructions to implement specific functions such as: I/O control logic, timing, counting, arithmetic, and data manipulation. A PC consists of a central processor, input/output interface, memory, and programming device which typically uses relay-equivalent symbols. The PC is purposely designed as an industrial control system which can perform functions equivalent to a relay panel or a wired solid-state logic control system.

Programming Environment: The total programming set-up that includes the interface, the languages, the editors and other programming tools.

PROLOG (Programming In Logic): A logic-oriented AI language developed in France and popular in Europe and Japan.

Proposition: A statement (in logic) that can be true or false.

Prototype: An initial model or system that is used as a base for constructing future models or systems.

Quality Assurance: Function of removing defective product from a manufacturing process.

Real Time: Taking place during the actual occurrence of an event. Real time refers to computer systems or programs that perform a computation during the actual time that a related physical process transpires, in order that the results of the computation can be recorded or used to guide the physical process. Examples of real-time systems include use of computers to guide airplane landings or to monitor assembly line processes.

Redundancy: Duplication of information or devices in order to improve reliability.

Reliability: The probability that a device will function without failure over a specified time period or amount of usage.

Robot: A mechanical device which can be programmed to perform some task of manipulation or locomotion under automatic control.

RS-232: EIA standard reflecting properties of serial communication link. RS-330: EIA standard governing closed-circuit television electrical signals. Specifies maximum amplitude of 1.0 V, peak to peak, including synchronization pulses.

Rule: A combination of facts, functions, and certainty factors in the form of a premise and an action. A pair, composed of an antecedent condition and a consequent proposition, that can support deductive processes such as backward-chaining and forward-chaining.

Rule-Based Program: A computer program that explicitly incorporates rules or ruleset components.

Rule-Based Systems: Computer systems in which information is encoded as rules rather than as algorithms or frames.

Rule-Interpreter: The control structure for a production rule system.

Scheduling: Developing a time sequence of things to be done.

Shell: A tool for building expert systems which is basically an empty system into which a knowledge base can be loaded.

Software: A computer program.

Solid-State Camera: see CCD and CID.

Speech Recognition: Recognition by a computer (primarily by pattern-matching) of spoken words or sentences.

Speech Synthesis: Developing spoken speech from text or other representations.

Speech Understanding: Speech perception by a computer.

Symbolic: Relating to the substitution of abstract representations (symbols) for concrete objects.

Symbolic Computing: This involves the processing of information in symbols instead of digits and characters. Allows for memory flexibility when adding new properties or values with a minimum of restructuring.

Symbolic Processing: A type of processing that primarily uses symbols rather tha numeric representations of data.

Theorem: A proposition, or statement, to be proved based on a given set of premises.

Theorem Proving: A problem-solving approach in which a hypothesized conclusion (theorem) is validated using deductive logic.

Throughput Rate: Generally refers to the number of objects to be examined per unit of time.

Time-Sharing: A computer environment in which multiple users can use the computer virtually simultaneously via a program that time-allocates the use of computer resources among the users in a near-optimum manner.

Top-Down Approach: An approach to problem-solving that is goal-directed or expectation-guided based on models or other knowledge. Sometimes referred to as "Hypothesize and Test."

Toolkit: A set of software tools for building an expert system, much more elaborate than a shell.

Tree Structure: A graph in which one node, the root, has no predecessor node, and all other nodes have exactly one predecessor. For a state space representation, the tree starts with a root node (representing the initial problem situation). Each of the new states that can be produced from this initial state by application of a single operator is represented by a successor node of the root node. Each successor node branches in a similar way until no further states can be generated or a solution is reached. Operators are represented by the directed arcs from the nodes to their successor nodes.

Uncertainty: A measure of how much confidence is placed in a piece of knowledge, used in uncertain reasoning.

Variable: A quantity or function that may assume any given value or set of values.

Verification: An activity providing qualitative assurance that a fabrication or assembly process was successfully completed.

Virtual Memory: A programming method that allows the operating system to provide essentially unlimited program address space.

Window: An application software design concept that (1) allows several programs to be run and displayed on the screen simultaneously, and (2) supports integration of data between application programs, e.g., spread sheet data displayed in one window may be included or merged with data stored in a word processing window. Use of multiple windows in a development environment permits system developers to monitor multiple processes or system states without the need to exit from one module to observe another.

Workstation: A system that (1) provides users with integrated, profession-specific functions, delivered through a single human interface (menu), (2) allows users to share work with other members of an organization by participating in a network of integrated functions located elsewhere in the organization, (3) has, at minimum, integrated graphics and word processing, plus terminal or file transfer communications, and (4) provides artificial intelligence developers with a single-user workstation.

World Knowledge: Knowledge about the world (or domain of interest).

World Model: A representation of the current situation.

Appendix 2

Commercially Available Software

1980/1000 WAVE MEASUREMENT LIBRARY

FUNCTION: Testing and Inspection

FOR: Minicomputer
HP1000

RTE-VI, RTE-A

PRICE: $1,000.00

Automated Technology Associates
7098 North Shadeland Avenue, Suite D-1
Indianapolis, IN 46220
(317) 842-9488

This licensed coversion of HP's 1980A Waveform Measurement Library for Series 200 Computers to the HP1000 Series is functionally compatible with the HP 1980A Product described in this catalog. The 1980/1000 Waveform Measurement Library links the HP 1980A/B Oscilloscope to HP 1000 computers and contains the software needed to characterize and compare time-domain wavefcrms automatically. ATA's 1980/1000 provides enhanced first-day measurement capabilities, and substantially reduces the program development time for specific Waveform Measurement applications. Waveform plotting capabilities are provided via ATA's APAPLOT/1000 products. ("Hooks" for automatic plotting via ATAPLOT are provided in the 1980/1000 software library modules.)1980/1000 Waveform Measurement Library for Series 200 is available to users in the multi-tasking HP1000 environment.

ACCEPTANCE SAMPLING-1000

FUNCTION: Process Control

FOR: Minicomputer
HP1000

RTE-IV

PRICE: $3,000.00

Hansford Data Systems, Inc.
3055 Brighton-Henrietta Town Line Road
Rochester, NY 14623
(716) 442-7110

Acceptance Sampling-1000 provides the capability to design and analyze lot-by-lot, single sampling plans for attributes based on hypergeometric, binominal, or Poisson probabilities. The package makes it possible to investigate plans for isolated lots arriving at the receiving dock, for lots from a continuous in-house process, and for other inspection circumstances. Lot quality is expressed as defectives, percent defective, or defects per hundred units, enabling analysis of MIL-STD-1050 plans.

AIDSTG

FUNCTION: Testing and Inspection

FOR: Mainframe
IBM, DEC/VAX

PRICE: Available upon request

Gateway Design Automation Corp.
235 Great Road
Littleton, MA 01460
(617) 486-9701

Automatic test vector generation for VLSI logic networks (50,000 gates and up) becomes practical with AIDS Test Generation (AIDSTG) system. AIDSTG makes test vector generation a push button operation for VLSI logic networks that employ the scan design discipline. Any one of four scan design disciplines may be used: 1) random access scan (Amdahl/Fujitsu), 2) scan set (Sperry-Univac), 3) level sensitive scan design (IBM), and 4) scan path (Nippon Electric).

ALSTAR DISTRIBUTE TEST SYSTEM (DTS)

FUNCTION: Testing and Inspection

FOR: Minicomputer
HP1000

RTE-VI

PRICE: $50,000.00

NCR Corporation-Wichita
3718 N. Rock Road
Wichita, KS 67226
(316) 688-8101

ATA-METER

FUNCTION: Testing and Inspection

FOR: Minicomputer
HP1000

9800 Native, RTE-VI, RTE-A

PRICE: $3,000.00

Automated Technology Associates
7098 N. Shadeland Avenue, Suite D-1
Indianapolis, IN 46220
(317) 842-9488

ATA-METER is a completely automated meter calibrator package for the generation and execution of calibration procedures for digital, analog and GPIB meters. As the first member of the ATACAL family of sophisticated automated calibration software, ATA-METER has unique features and construction to insure that your software investment is protected from obsolescence due to hardware or calibration requirements changes.

ATLAS-10 INTERACTIVE TEST PROGRAM DEVELOPMENT SYSTEM

FUNCTION: Testing and Inspection

FOR: Minicomputer
HP1000

RTE-IV, RTE-A

PRICE: $25,000.00

Lexico Enterprises, Inc.
8757 Georgia Avenue, Suite 545
Silver Spring, MD 20910
(301) 588-0400

ATLAS-10 is a combined text editor and syntax verifier and recognizes pure IEEE 416 ATLAS language. It is designed for use by test engineers in the development of test procedures/programs in accordance with ATA 100 and MIL STD TRD requirements.

ATLAS/1000

FUNCTION: Testing and Inspection

FOR: Minicomputer
HP1000

RTE-IV

PRICE: $45,000.00

Lexico Enterprises, Inc.
8757 Georgia Avenue, Suite 545
Silver Spring, MD 20910
(301) 588-0400

ATLAS (Abbreviated Test Language for All Systems) is an English-like, high-level language for describing electronic equipment test procedures. ATLAS statements describe actions to be accomplished in an unambiguous, self-documenting format which can be translated by a computer system into instructions that control automatic test equipment (ATE). The ATLAS language structure is defined in ANSI/IEEE 416.

CALIBRATION-1000

FUNCTION: Testing and Inspection

FOR: Minicomputer
HP1000

RTE-IV

PRICE: $5,000.00

Hansford Data Systems, Inc.
3055 Brighton-Henrietta Town Line Road
Rochester, NY 14623
(716) 442-7110

Calibration-1000 is a comprehensive calibration scheduling and control system for gages and test instruments. This system is designed to reduce manufacturing costs and improve productivity.

CALIBRATION-98X6

FUNCTION: Testing and Inspection

FOR: Minicomputer
Series 200

9800 Native

PRICE: $2,500.00

Hansford Data Systems, Inc.
3055 Brighton-Henrietta Town Line Road
Rochester, NY 14623
(716) 442-7110

CALIBRATION-98X6 is a comprehensive calibration scheduling and control system for gages and test instruments. This system is designed to reduce manufacturing costs and improve productivity.

CHECKPOINT LABEL INSPECTOR

FUNCTION: Machine Vision and Image Processing, Testing and Inspection, Food Industries

FOR:
Sold only with Checkpoint Processor

PRICE: Available upon request

Cognex Corp.
72 River Park Street
Needham, MA 02194
(617) 449-6030

When equipped with label inspection software, Checkpoint can detect missing, multiple, and skewed labels visually, labels that are incorrect in size, are improperly positioned, or have major wrinkles or folds.

CHECKPOINT PCB

FUNCTION: Machine Vision and Image Processing, Testing and Inspection. Electronics

FOR:
Sold only with Checkpoint Processor

PRICE: Available upon request

Cognex Corp.
72 River Park Street
Needham, MA 02194
(617) 449-6030

When equipped with PCB software, Checkpoint can inspect populated printed circuit boards visually for proper lead placement and crimp, missing components, wrong components, damaged components, and improperly oriented components. Checkpoint can also inspect the quality of printing on IC components, and can read circuit board numbers for automatic initiation of electronic board test routines.

CHECKPOINT PQI (PRINT QUALITY INSPECTION)

FUNCTION: Machine Vision and Image Processing, Testing, and Inspection, Printing & Publishing

FOR:
Sold only with Checkpoint Processor

PRICE: Available upon request

Cognex Corp.
72 River Park Street
Needham, MA 02194
(617) 449-6030

When equipped with PQI software, Checkpoint can inspect the output of printers, typewriters, copiers, and other equipment visually. The system uses artificial intelligence techniques to identify those defects most objectionable to humans. Among the attributes CHECKPOINT measures are: shape, contrast, angle, scale, elevation, kerning, line pitch, interline spacing, left margin deviation, and the extent to which a liftoff character has been removed.

CHI

FUNCTION: CAD/CAM, Expert System Development, Testing and Inspection

FOR: Mainframe
Lisp Machines

PRICE: Available upon request

Kestrel Institute
1801 Page Mill Road
Palo Alto, CA 94304
(415) 493-6871

CHI is a knowledged-based software synthesis system that combines advanced expert system, compiler, and data base technology. CHI supports rapid prototyping as well as synthesis of efficient programs from very high level specifications. CHI's programming knowledge includes knowledge on data structures, algorithm design, and optimization. CHI further includes sophisticated program (specification) analysis and compilation tools. CHI can be tuned toward specific applications by adding domain-specific knowledge. CHI should be particularly suited for building expert systems and software generators.

CMM QUALITY MANAGEMENT—1000 FOR ZEISS

FUNCTION: Testing and Inspection

FOR: Minicomputer
HP1000

RTE-IV

PRICE: $10,000.00

Hansford Data Systems, Inc.
3055 Brighton-Henrietta Town Line Road
Rochester, NY 14623
(716) 442-7110

The Coordinate Measuring Machine (CMM) Quality Management—1000 for Zeiss package provides on-line data collection, storage, and management capabilities, creating a historical parts management data base.

CPA/1000 CYLINDER PRESSURE ANALYZER

FUNCTION: Testing and Inspection

FOR: Minicomputer
HP1000

RTE-IV

PRICE: $69,500.00

Dayton Scientific, Inc.
92 Westpark Court
Dayton, OH 45459
(513) 433-9600

The CPA/1000 system is designed for the specialized task of internal combustion engine cylinder pressure analysis. Unique data acquisition hardware provides high data rate and simultaneous sampling of up to sixteen analog input channels. These input channels normally acquire the outputs of charge amplifiers used with piezoelectric cylinder pressure transducers. Samples are acquired in response to pulses from a crankshaft position encoder at angular intervals of ½ or 1 of rotation. The maximum data acquisition rate for any one channel is 72,000 samples/second. The maximum continuous data rate to disc storage is 250,000 samples/second, which sets an upper limit on the combined data rate from all active channels.

DPG/LINK DIGITAL PROGRAM GENERATORS

FUNCTION: Testing and Inspection

FOR: Minicomputer
HP1000

RTE-IV, RTE-VI

PRICE: $7,000.00

Lexico Enterprises, Inc.
8757 Georgia Ave., Suite 545
Silver Spring, MD 20910
(301) 588-0400

DPG/LINK is hosted on an HP 1000 computer, requires HP Test-Aid III software, and is usable only in conjunction with the HP 3062A series of upgraded In-Circuit Testers with the Series 200 HP 9826/36 desktop controller.

EXEC-1000 TEST EXECUTION/SCHEDULING SYSTEM

FUNCTION: Testing and Inspection

FOR: Minicomputer
HP1000

RTE-IV, RTE-VI, RTE-A

PRICE: $25,000.00

Lexico Enterprises, Inc.
8757 Georgia Ave., Suite 545
Silver Spring, MD 20910
(301) 588-0400

This software package provides for a standard test operator interface, test scheduling, execution and data logging. EXEC-1000 is written in FORTRAN, and utilizes the IMAGE 1000 data base utilities (available from Hewlett-Packard Co.). EXEC-1000 includes a fault isolation routine table look-up procedure which examines failed measurements and determines the most probable cause. Minimum retest requirements are enforced during subsequent testing to ensure that all failures are detected/corrected.

HP 19800A WAVE FORM MEASUREMENT LIBRARY

FUNCTION: Testing and Inspection

FOR: Minicomputer
Series 200

9800 Native

PRICE: $1,000.00

Hewlett-Packard
19447 Pruneridge Avenue
Cupertino, CA 95014
(408) 725-8111

Linking the HP 1980A/B oscilloscope measurement system to HP computers, the HP Model 19800A Waveform Measurement Library contains the software needed to characterize and compare time-domain waveforms automatically. The HP 19800A can make measurements on the first day and substantially reduces the time needed to develop software for specific applications.

HP 85015A SYSTEM SOFTWARE

FUNCTION: Testing and Inspection

FOR: Microcomputer
HP 9816A, HP 9826A, HP 9836A, HP 9836C, HP 9920A

Basic 2.0 with 750 KBytes RAM

PRICE: $1,500.00

Hewlett-Packard Co.
P.O. Box 10301
Palo Alto, CA 94303-0890
(707) 525-1400

The HP 85015A System Software is designed to help make microwave scalar measurements more productive and help operate HP 8756S Automatic Scalar Network Analyzer System. At the heart of the 8756S System is the HP 8756A Scalar Network Analyzer, a versatile instrument that measures reflection and transmission parameters and displays them on two independent channels. With different test sets and high-directivity directional bridges, the 8756A covers a frequency range of 10 MHz to 40 gHz and beyond. The signal source in the 8756S System is an HP 8350B Sweep Oscillator with an RF plug-in, or an 8340A Synthesized Sweeper. Both sweep oscillators, like the 8756A, are fully programmable via a HP-IB (Hewlett-Packard Interface Bus). System control is provided by the powerful Series 200 computer (HP 9836A/26A/16A). The system software and computer permit this scalar system to make accurate reflection and transmission measurements with impressive speed, convenience, and thoroughness. The key features of this computer aided testing software are: simple menu operation, easily customized measurements, flexible plotting and printing formats, and real-time limit testing.

HP 85016A TRANSMISSION LINE TEST SOFTWARE

FUNCTION: Testing and Inspection

FOR: Micrcomputer
HP 9816A, HP 9826A, HP 9836A, HP 9836C,
HP 9920A

Basic 2.0 with 1 MByte RAM

PRICE: $4,500.00

Hewlett-Packard Co.
P.O. Box 10301
Palo Alto, CA 94303-0890
(707) 525-1400

The HP 85016A transmission line test software adds fault location capability to the HP 8756S scalar network analyzer system. This software allows the customer to characterize a coaxial cable or a waveguide run completely by measuring the insertion loss, return loss and distance to fault at the transmission line's operating frequency (from 10 MHz to 18 MHz). The HP 85016A has a sweep oscillator, a directional bridge and one or more diode detectors. The software measures the frequency response of the transmission line, computes an inverse Fast Fourier Transform (FFT) and then plots return loss as a function of distance along the line. The customer can locate bad connections and faults in transmission lines up to 1000 feet long. The key features of the HP 85016A transmission line test software are

- Speed—it can calculate and display the FFT in 20 seconds.
- Store Test Setups and Data—easily stored results on disk.
- Three User Modes combine ease-of-use with versatility.
- Selectable Resolution and Range—resolution can be 0.25%, 0.5%, 1%, or 2% or range
- Correction for Multiple Faults and Wave Guide Dispersion

HP 85864A EMI MEASUREMENT SOFTWARE

FUNCTION: Testing and Inspection

FOR: Microcomputer
Hewlett-Packard

PRICE: $3,020.00

Hewlett-Packard
1820 Embarcadero Road
Palo Alto, CA 94303
(707) 525-1400

HP 85864A EMI Measurement Software performs MIL-STD and commercial emissions tests using an HP 8568A/B spectrum analyzer, HP 9000 Model 226 or 236 desktop computer, and various accessories.

HP QUALITY DECISION MANAGEMENT/1000

FUNCTION: Testing and Inspection

FOR: Minicomputer
HP1000

RTE-VI

PRICE: $40,000.00

Hewlett-Packard Company
19447 Pruneridge Avenue
Cupertino, CA 95014
(408) 725-8111

HP Quality Decision Management/1000 is an applications software package for analyzing manufacturing process and product quality. The package provides control charts that help production and quality assurance engineers identify and prioritize statistically significant product defects and manufacturing process problems. Process and production engineering departments can use data collected on-line to generate scattergrams, histograms, and detailed causes of product quality deviations.

HP QUALITY DECISION MANAGEMENT/1000-92121A

FUNCTION: Testing and Inspection

FOR: Minicomputer
HP 1000

VC+, RTE-A

PRICE: $25,000.00-$45,000.00

Hewlett-Packard
370 W. Trimble Road
San Jose, CA 95131
(408) 263-7500

HP Quality Decision Management/1000 is an applications software package for analyzing manufacturing processes and product quality. The package provides control charts and Pareto charts that help production and quality assurance engineers identify and prioritize statistically significant product defects and manufacturing process problems. Engineering departments can use data collected online to generate scattergrams, histograms, and tabular reports. These outputs identify the manufacturing process causes of product quality deviations. A menu and prompt/response approach allows engineers without programming experience to configure data collection transactions, specify report and graph formats, archive data, and perform system maintenance functions. Extensive "hooks" for user programs provide additional data input, output, and analysis flexibility.

INSTRUCTION MANAGEMENT-3000

FUNCTION: Testing and Inspection

FOR: Minicomputer
HP3000

MPE

PRICE: $10,000.00

Hansford Data Systems, Inc.
3055 Brighton-Henrietta Town Line Road
Rochester, NY 14623
(716) 442-7110

Instruction Management-3000 provides a system for online storage and updating of work instructions for Receiving, Sampling, Inspection, Assembly, Calibration, Maintenance, and other manufacturing functions. Files are created and edited for specification and procedure changes. They are immediately available to all pertinent personnel.

LASER MACHINE TOOL CALIBRATION SYSTEM

FUNCTION: Testing and Inspection

FOR: Microcomputer
Series 80

80 Native

PRICE: Available upon request

Weldon Machine Tool, Inc.
1800 W. King Street
York, PA 17404
(717) 846-4000

The Laser Machine Tool Calibration System was designed by Weldon Machine Tool to facilitate the use of Hewlett-Packard's 5505A and 5508A Laser Interferometer Measuring Systems in the calibration of machine tools. It is an ideal system for both manufacturers and users of machine tools to measure accuracy, repeatability, backlash, straightness, squareness, pitch, and yaw for up to 24 axes and document in both printed and chart form the complete results with statistical 6 sigma limits, all done with a technician whose only training is the mechanical setup of the laser interferometer system.

MAMA: MEASUREMENTS & MICROWAVE ANALYSIS

FUNCTION: Testing and Inspection

FOR: Minicomputer
Series 200

9800 Native

PRICE: $5,000.00

Made-It Associates, Inc.
21 Daniel Drive
Burlington, MA 01803
(617) 272-6992

MAMA is a calibration, control and analysis program for use with the HP-8409 Microwave Automatic Network Analyzer. MAMA calibrates the HP-8409 ANA, measures all S-parameters on 1- and 2-port devices, and converts frequency-domain data into time domain. Interactive graphics allow easy analysis of elements and location of discontinuities. Analysis data may be printed or plotted. Over 20 parameters may be viewed, including rectangular plots of magnitude, angle, VSWR, gain, and return loss. Polar plots of input and output impedance are also available.

MICRO/QC

FUNCTION: Quality Control

FOR: Microcomputer
8 bit CP/M, IBM Compatibles 16 bit

CP/M, MS DOS

PRICE: $500.00

Generates process quality control charts from either data files downloaded from mainframes or from manual input. It features X, R, Sigma, P, NP, U and C control charts and several statistical analysis programs including Analysis of Menus for single and multiple variable situations. Manuscript text editor is included as a part of this package for doing manual data entry.

Stachos, Inc.
14 College Street
Schenectady, NY 12305
(518) 372-5426

OPTICAL RADIATION MEASUREMENT SOFTWARE

FUNCTION: Testing and Inspection

FOR: Microcomputer
Series 80

80 Native

PRICE: $795.00-$1,225.00

Optronic Laboratories offers a series of software packages which are designed to make a variety of automatic optical radiation measurements using the Models 740 (740A-D), 742 and 746 (746-D) spectroradiometer systems. The versatility and flexibility designed into these spectroradiometers has also been designed into the software packages.

Optronic Laboratories, Inc.
730 Central Florida Pkwy.
Orlando, FL 32824
(305) 857-9000

PMS

FUNCTION: Testing and Inspection

FOR: Minicomputer
LSI 11

TSX

PRICE: Depends on configuration

Spatial Data Systems
420 S. Fairview
Santa Barbara, CA 93116
(809) 967-2383

Controls parts measurement system. Complete systems includes a software controlled moving table (x and y axis), camera, imager, measuring and calibration software, and statistics storage.

PRISM

FUNCTION: Testing and Inspection

FOR: Minicomputer
Series 200

9800 Native

PRICE: $6,000.00

Anatrol Corporation
10895 Indeco Drive
Cincinnati, OH 45241
(513) 793-8844

PRISM is an advanced model analysis program which also provides spectrum data storage, plotting, and manipulation capabilities for 2-channel fourier analyzers. The speed and power of PRISM is due in large part to the fact that it is written in Pascal for the Series 200 desktop computers. With all of its advanced capabilities, it has been made easy to operate by the inclusion of extensive self-help features.

PROCESS ANALYSIS-1000

FUNCTION: Testing and Inspection

FOR: Minicomputer
HP1000

RTE-IV, RTE-VI

PRICE: $10,000.00

Hansford Data Systems, Inc.
3055 Brighton-Henrietta Town Line Road
Rochester, NY 14623
(716) 442-7110

Process Analysis-1000 provides numeric and graphic analytical tools for many statistical quality control applications to increase productivity and reduce manufacturing costs.

PROCESS ANALYSIS-98X6

FUNCTION: Testing and Inspection

FOR: Minicomputer
Series 200

9800 Native

PRICE: $3,500.00

Hansford Data System, Inc.
3055 Brighton-Henrietta Town Line Road
Rochester, NY 14623
(716) 442-7110

Process Analysis-98X6 provides comprehensive data storage, data management, and analytical tools to increase productivity and reduce manufacturing costs. The system is designed for use by quality engineers and technicians. Graphic displays enable interpretation and trouble-shooting of production problems.

PRODUCTION QUALITY MANAGEMENT-1000

FUNCTION: Testing and Inspection

FOR: Minicomputer
HP1000

RTE-IV, RTE-VI

PRICE: $10,000.00

Hansford Data Systems, Inc.
3055 Brighton-Henrietta Town Line Road
Rochester, NY 14623
(716) 442-7110

The Production Quality Management-1000 package provides on-line storage and complete data management capabilities for variables data. Creating and maintaining a data base is accomplished without programming.

PRODUCTION QUALITY MANAGEMENT-3000

FUNCTION: Testing and Inspection

FOR: Minicomputer
HP3000

MPE

PRICE: $10,000.00

Hansford Data Systems, Inc.
3055 Brighton-Henrietta Town Line Road
Rochester, NY 14623
(716) 442-7110

The Production Quality Management-3000 package provides online storage and complete data management capabilities for variables data. Creating and maintaining a data base is accomplished without programming.

QC2 QUALITY CONTROL/ASSURANCE INFORMATION SYSTEM

FUNCTION: Testing and Inspection

FOR: Minicomputer
HP3000

MPE

PRICE: Available upon request

Hutchison Lamb & Associates, Inc.
813 Ridge Lake Blvd., Suite 310
Memphis, TN 38119
(901) 767-0831

QC2 is an online, interactive computer system designed to assist quality control laboratories which must sample, test and evaluate multiple types of products and materials using various testing laboratories and procedures. QC2 encompasses all of the sampling, testing, evaluating, test scheduling, and reporting functions within the quality control departments of drug, cosmetic, and food manufacturers.

QMS (QUALITY & MANUFACTURING SYSTEM)

FUNCTION: Testing and Inspection

FOR: Minicomputer
HP3000

MPE

PRICE: $35,000.00-$135,000.00

QMS is a comprehensive software product for quality assurance and materials/usage management. Because QMS is parameter driven, it can be adapted to fit many types of environments. QMS currently consists of four totally integrated subsystems which span eight data bases.The four subsystems are quality and specifications, order processing, materials, and production management reporting.

SATCOM
4521 Professional Circle
Virginia Beach, VA 23455
(804) 499-9803

QMS PROGRAMS

FUNCTION: Testing and Inspection

FOR: Minicomputer
PDP DEC, VAX

RSX

PRICE: $8,000.00-$45,000.00

QMS Programs are an integrated set of computer program modules that manage quality control information. QMS Programs provide for the on-line collection, documentation, storage, reporting, and analysis of quality test and inspection results. Features include the PROTOCOL dialog, which allows users to custom-tailor any number of QC applications within a single system; data entry from a variety of devices, including a compatible automatic screen generation module, EZ DATA; interactive reporting and/or statistical analysis capabilities in a variety of user-controlled formats; online help messages; simulation capabilities that allow "what-if" games with new-product startups; and built-in graphics.

John A. Keane and Associates
20 Nassau Street
Princeton, NJ 08540
(609) 924-7904

QUALITY MANAGEMENT AND PRODUCTION (QMAP)

FUNCTION: Testing and Inspection

FOR: Minicomputer
HP3000

MPE

PRICE: $30,000.00-$150,000.00

BSE, Inc.
77 West Main Street
Freehold, NJ 07728
(201) 780-2626

MAP is dedicated to batch flow and continuous flow manufacturers who integrate quality attributes to "lots" or units of production or finished inventories. The system automatically controls stock selection to production and/or customer orders. Full birth-to-death accountability and movement tracking are an integral feature of the software. Plants with a need to monitor quality on materials (raw/in process/finished goods) and to check test results against both product and customer specs need this type of system. However, plants with complex product control requirements and large, high dollar material batches/lots achieve the highest cost benefits from system use. The system can handle quality oriented tests.

SCREENDUMP (HP8566/68 SCREEN HARD COPY)

FUNCTION: Testing and Inspection

FOR: Minicomputer
Series 200, HP1000

9800 Native, RTE-VI, RTE-A

PRICE: $495.00

Automated Technology Associates
7098 North Shadeland Ave., Suite D-1
Indianapolis, IN 46220
(317) 842-9488

SCREENDUMP is an aid for rapid documentation and communication of the test results obtained with the HP8566 and 8568 Spectrum Analyzers. Hard copy of these screen displays are just as important to the understanding of test reports and documentation as the screen display itself is to the user of spectrum analyzer. SCREENDUMP provides a friendly utility for use with the Spectrum Analyzer and an intelligent controller to dump the full contents of the Spectrum Analyzer screen to any supported graphics peripheral device such as pen plotters, or the controller screen. (Dot matrix printers may be accessed from the HP9835 with the binary plotting package.) The hard copy includes all labels, cursor positioning, and all graphics displayed on the screen at the time SCREENDUMP is invoked. The hard copy can be the full size of the dump device and uses the full resolution of the spectrum analyzer screen.

SYS-FLO (THE PROCESS PIPING DESIGN AND ANALYSIS PROGRAM)

FUNCTION: Energy and Facility Management, Testing and Inspection, Food Industries, Paper

FOR: Microcomputer
Any microcomputer running CP/M, MS DOS, or PC DOS

CP/M, CP/M+, CP/M 86, PC DOS, MS DOS

PRICE: $750.00

Engineered Software
2420 Wedgewood Dr.
Olympia, WA 98501
(206) 786-8545

SYS-FLO is used to design and analyze process piping systems. The SYS-FLO Pipe Editor is used to establish the piping system pipeline, and component interconnections. The pipelines can be connected into series and series/parallel flow paths. Pipeline components such as pumps, filters, heat exchangers can be modeled, tested, and then installed into the system. The system configuration can be edited or modified by the pipe editor. The only limit to the size of process system the program can analyze is the available diskette space. Printed reports provide the pressures at each point in the system and the flow rate in the parallel flow paths. The printed reports provide the final results and enough intermediate results to verify the solution.

TERANET ATE NETWORKING SYSTEM

FUNCTION: Testing and Inspection

FOR: Minicomputer
Pr ' DEC

RSX

PRICE: $61,500.00-$75,000.00

Teradyne, Inc.
183 Essex Street
Boston, MA 02111
(617) 482-2700

The Teradyne ATE Networking System (TERANET) is a hierarchical networking system for Teradyne's broad line of ATE and laser adjust systems. The TERANET System consists of three major elements: The Test System Director (TSD); a multidrop communications link and interfaces have a data transfer rate of 1 million bits per second; software tools that perform a variety of test department support functions. Software for test floor supervision collects, stores, and displays, in real time, significant network events and lets executive software be stored centrally on disk. Software for remote file transfer supports bidirectional file transfer between the TSD and a higher-level computer outside the test department. This is helpful in factory communications.

TMGR (TEST MANAGER)

FUNCTION: Testing and Inspection

FOR: Minicomputer

PRICE: N/A

General Electric Company
P.O. Box 8
Schenectady, NY 12301
(518) 385-8641

TMGR, or Test Manager, is an extensive software package that makes it possible to automate the testing of many types of heavy industrial equipment almost completely. This new family of programs is at the heart of several computer systems GE has assembled for its internal use. In its various applications, TMGR handles all the routine test management tasks such as issuing the computer-language "orders" required to call up remotely operated sensors and retrieve, store, and display data from them. Unlike conventional computerized data-acquisition systems, however, it is also able to perform detailed engineering analyses of the data it receives while the test is going on.

TUTSIM FULL PROGRAM

FUNCTION: CAD/CAM, Manufacturing Machinery Control, Process Control, Robotics, Energy and Facility Management, Testing and Inspection

FOR: Minicomputer, Microcomputer
IBM PC/XT, Apple II series, PDP-11, Osborne, Kaypro II

DOS 2.0, Apple DOS, CP/M80, CP/M 5½" for Osborne & Kaypro II

PRICE: $475.00–$525.00

Applied i
200 California Avenue #214
Palo Alto, CA 94306
(415) 325-4800

TUTSIM, the program for simulation of continuous dynamic systems, provides Micro and Mini users with the opportunity to model the processes of any system which can be described by differential equations. TUTSIM is a block diagram oriented language which produces both numeric and graphic output. Since the block diagrams describe mathematical relationships, TUTSIM is interdisciplinary, with applications in numerous engineering fields, physiological and biological model studies, chemical system studies, econometric, and sociological system models. The TUTSIM manual instructs the user on modeling a system by translating the terms of the equations which describe that system into block diagram form or by directly assembling blocks selected from the 36 block elements supplied by TUTSIM. Inputting the system model into the program is accomplished by entering block structural and variable parameters in single line format. As a simulation is run, output from any block may be graphed. Parameters may be changed by simple one line commands, and simulation rerun. Graphic results can be overlaid, thus making TUTSIM an invaluable tool for optimizing systems. The professional version of this program for microcomputers handles up to 999 blocks and sells for $475.00 to $525.000.

TUTSIM SHORT FORM PROGRAM

FUNCTION: CAD/CAM, Manufacturing Machinery Control, Process Control, Energy and Facility Management, Testing and Inspection

FOR: Microcomputer
IBM PC/XT, Apple II series, Osborne I, Kaypro II
CP/M 80 (8″)

DOS 2.0, Apple DOS, CP/M80, CP/M 5½″ for
Osborne I & Kaypro II

PRICE: $29.95

Applied i
200 California Avenue #214
Palo Alto, CA 94306
(415) 325-4800

TUTSIM, the program for simulation of continuous dynamic systems, is not available in a "short form" suitable for evaluation, demonstration, and instructional study. The "short form" has the same syntax, the same manual, an extended set of examples on the disk, and all the other features of the professional version of the program but is limited to ten blocks. Thus it is ideal for self instruction in the art of simulation and for evaluation of the TUTSIM program for a given user's needs. Program description is the same as for the Full Program.

VARIABLES ANALYSIS-3000

FUNCTION: Testing and Inspection

FOR: Minicomputer
HP3000

MPE

PRICE: $10,250.00

Hansford Data Systems, Inc.
3055 Brighton-Henrietta Town Line Road
Rochester, NY 14623
(716) 442-7110

Variables Analysis-3000 provides numeric and graphic analytical tools for many statistical quality control applications to increase productivity and reduce manufacturing costs.

Index

AcousTechs system, 61
Adaptive Technologies, Inc., 170
Allen-Bradley, 147, 186
Allis Chalmers, 27
American Channels, 32
American National Standards
    Institute, 91
Applied Intelligent Systems, 45,
    111, 133
Arthur D. Little, Inc., 167
Artificial intelligence program-
    ming languages, 172
ASEA, 82
AT&T, 189
AUTOFACT, 4
Automated Industries Division
    of Sperry, 59
Automated Intelligence, Inc., 136

Automated Vision Association,
    46
Automatic inspection, 65, 108
Automobile body gauging, 80
Automatix, 4, 118
Automatrix Autovision, 123

Barry Wright Corp., 4
Bath, University of, 124
Beckman Instruments, Inc.,
    168
Bell Laboratories, 34
Bergishe Stahl Industries,
    85
Binary processing, 43
Bin picking, 3
Boeing, 147, 151
BW Electronics, 65

CAM-I, 22, 169
Carnegie Mellon University,
    45
Casting flaw detection, 123
Cellular processing, 45
Charged-coupled devices, 35,
    121
Checkpoint 1100, 115
Cochlea Corp., 4, 63
Cognex Corp., 115, 132
Component Evaluation, 167
Computer-aided design, 11, 21
Computer-aided engineering, 11
Computer-aided manufacturing,
    21
Computer-aided process plan-
    ning, 30
Computer communications, 146
Computer-integrated manufac-
    turing, 11, 12, 15, 17, 21,
    189, 191
Computer simulation, 153
Concord Data Systems, 147
Concurrent National Dept. of
    Energy Welding Systems,
    59
Connectivity analysis, 39
CONSIGHT vision system, 3, 4,
    39
Control Automation, 4
Conval Corp., 84
Coordinate measuring machines,
    106
Corning Glass, 168

DataMan, 132
D. Appleton Co., Inc., 140
Deere and Co., 17, 19, 190
Detector alarm check, 86
Diffracto, Ltd., 111

Digital Equipment Corp., 140,
    147
Dupont, 147

EG&G Idaho, 59
Eisai Co., Ltd., 65
Electrical contact testing, 88
Electrical Laboratory, 170
Environmental Research Institute
    of Michigan, 142
EOIS, 133
Everett/Charles robot, 89
Expert systems, 163
    for quality assurance, 167
    technology of, 171
    tools for building, 182

Factory of the future, 12, 15
Federal Drug Administration, 65
Filtec systems, 65
Flexible manufacturing systems,
    154, 159
Ford, 147
Frames, 176
Frost, Inc., 192

General Electric, 133, 145, 147,
    159, 193
General Motors, 17, 26, 100,
    146, 147, 198
Gould Modicon, 147
GRASP, 155
Gray scale, 43

HEI Corp., 154
Hewlett-Packard, 147

IBM, 4, 27, 82, 121, 147
IEEE, 147
Industrial Dynamics Co., Ltd., 65

Industrial robot, 2
Ingersoll Milling Machine Co.,
    17
Inspection, 127
    alphanumeric character, 131
    automated closure, 134
    container and label, 128
    pharmaceutical product count-
        ing, 136
Integrated Automation, 121
Intel, 147
Intelledex, 4, 117

Jet Propulsion Lab, 43
Just-in-time, 9, 12, 25

Key Image, 133
Knowledge-based system, *see*
    Expert system
Knowledge engineering, 166
Knowledge representation, 173
Kodak, 100, 147

Lakso Reformer, 111, 113
Laser Probe, 111
Lighting, 36
Lockheed-Georgie, 31
Lord Corporation, 4

MIT, 4, 43
Machine Intelligence Corp., 3
Machine vision, 33-48
    implementing, 141
Machine Vision International,
    45
Mailbox approach, 24
Manufacturing Automated Pro-
    tocol, 147, 199, 206
    seven layers of, 149
    technology of, 148

Manufacturing Resource Plan-
    ning, 11
Mathematical morography, 45
McDonnell Aircraft Co., 17
McDonnell Douglas, 170
Merck & Co., Inc., 200
Michigan, University of, 45
Midus controller, 59
Minnesota Mining & Manufac-
    turing (3M), 204
Motorola, 147
MRP, 27, 28

National Aeronautics and Space
    Administration, 17
National Bureau of Standards,
    4, 22, 124, 147
National Computer Conference,
    147
National Research Council, 17
National Science Foundation, 2
Numerical Control, 11

Oakland Engineering, 61
Object Recognition Systems,
    Inc., 124
Operating system, 92
OptiScan 4 system, 65

Pattern Processing Technologies,
    Inc., 111
Pharmaceutical inspection, 111
Poughkeepsie Robotic Tester, 82
Proctor and Gamble, 124
Product Labels, 167
Product Safety, 167
Production rules, 174
Production systems, 173
PROM, 59
PUMA robot, 3

Purdue University, 2, 3, 24, 169

Reis robot, 85
Renault, 147
Resolution, 36
Rhode Island, University of, 2
Robotic inspection, 71-89
  implementing, 143
Robots, 73
  continuous path, 73
  computer-controlled, 73
  cylindrical, 74
  intelligent, 76
  jointed-spherical, 74
  programmable, 73
  programming languages, 74
  rectangular, 74
  reliability, 76
  sensory, 73
  spherical, 74
  vision, 76
Rockwell, 59
Rule interpreter, 175

Saab, 101
Sargent & Co., 27
ScanSystem Model 200, 124
Search, 178
  techniques, 179
Seiko, 88
Selecom Optocator, 80
Selective Electronic, Inc., 80
Semantic networks, 176
Semiconductors, 114
Sensors, 49-69
Sheet metal inspection, 102
Software, 91-97
  how to select, 93
  reliability, 95
Solid-state camera, 34, 35

Sonovision system, 63
Sperry Products Division, 59
Square D, 88
SRI International, 2, 39
Stanford University, 2, 3, 20, 163
Star approach, 24
Strategy and Implementation
    Program, 140
Synthetic Vision Systems, 45, 114
Systems operation, 175

TCM board tester, 82
Texas Instruments, 202
Teknowledge, Inc., 167
Three-dimensional vision, 43
Tunnell, K. W., Co., Inc., 94

Ultrasonic inspection, 58, 63
Ultrasonic leak detection, 61
U.S. Air Force, 22
U.S. Department of Defense
    DARPA, 47, 51
U.S. Department of Energy, 59

Valve testing, 82
Video Tek, Inc., 135
VLSI, 47
Volkswagen, 147

WAVE programming system, 3
Webspec 3000, 121
Weld inspection with ultrasonics,
    58
Westinghouse Defense and Elec-
    tronics Center, 17
Whirlpool, 207

Xerox Network Services, 152
X-ray testing, 85

ZapatA Industries, Inc., 134